FLUID POWER CONTROL

Hydraulics and Pneumatics

Ahmed Abu Hanieh

Viva Books
New Delhi | Mumbai | Chennai | Kolkata
Bengaluru | Hyderabad | Kochi | Guwahati

First Indian Edition 2012
Reprinted 2015

VIVA BOOKS PRIVATE LIMITED

- 4737/23 Ansari Road, Daryaganj, New Delhi 110 002
  E-mail: vivadelhi@vivagroupindia.net, Tel. 42242200
- 76, Service Industries, Shirvane, Sector 1, Nerul, Navi Mumbai 400 706
  E-mail: vivamumbai@vivagroupindia.net, Tel. 27721273, 27721274
- Megh Towers, Old No. 307, New No. 165, Poonamallee High Road, Maduravoyal
  Chennai - 600 095 E-mail: vivachennai@vivagroupindia.net, Tel. 23780991, 23780992
- B-103, Jindal Towers, 21/1A/3 Darga Road, Kolkata 700 017
  E-mail: vivakolkata@vivagroupindia.net, Tel. 22836381, 22816713
- 7, Sovereign Park Apartments, 56-58, K. R. Road, Basavanagudi, Bengaluru 560 004
  E-mail: vivabangalore@vivagroupindia.net, Tel. 26607409, 26607410
- 101-102 Mughal Marc Apartments, 3-4-637 to 641, Narayanguda, Hyderabad 500 029
  E-mail: vivahyderabad@vivagroupindia.net, Tel. 27564481, 27564482
- Beevi Towers, First Floor, SRM Road, Kaloor, Kochi 682 018, Kerala
  E-mail: vivakochi@vivagroupindia.net, Tel: 0484-2403055, 2403056
- 232, GNB Road, Beside UCO Bank, Silpukhuri, Guwahati 781 003
  E-mail: vivaguwahati@vivagroupindia.net, Tel: 0361-2666386

www.vivagroupindia.com

Published by arrangement with:

Cambridge International Science Publishing
7, Meadow Walk, Great Abington
Cambridge CB1 6AZ, UK

ISBN : 978-81-309-1988-1

Published by Vinod Vasishtha for Viva Books Private Limited, 4737/23, Ansari Road, Daryaganj, New Delhi - 110 002. Printed and bound by Sanat Printers, Kundli.

Acknowledgement

With the warmest feelings of love, I express my gratitude to my wife and children who were the strongest support for me to work on this book with their patience and understanding.

About the author

Ahmed Abu Hanieh

The author has obtained his bachelor degree in Mechanical Engineering from Birzeit University (BZU) in Palestine in 1993, the Masters and PhD degrees from Universite Libre de Brussels (ULB) in Brussels in 2003. He worked as an assistant professor in Birzeit University in 2004 and was assigned as a chairperson of the Mechanical Engineering Department in 2008. Abu Hanieh is the author of some journal papers and international conference proceedings in different disciplines of mechatronic systems and participated in establishing specialized local and international associations.

Preface

The idea of this book came out after five years of teaching the fluid power control course in Birzeit University besides to the long experience of the author in the real hydraulic and pneumatic systems which is reflected on the different applications of the discussed circuits. Most of the existing books in this field discuss the hydraulic and pneumatic systems in concentrating on the design and components of the system without going deep enough into the problem of dynamic modelling and control of these systems. This book attempts to compromise between theoretical modelling and practical understanding of fluid power systems by using modern control theory based on implementing Newton's second law in second order differential equations transformed into direct relationships between inputs and outputs via transfer functions or state space approach.

The first chapter begins with a fluid mechanics background that discusses the basic principles of Pascal and Bernoulli necessary for the fluid power calculations taking into account the pressure losses due to friction effect in pipes and fittings. Chapter two handles the different techniques of modelling used to represent the linear and rotary actuators, the control valves and the hydrostatic transmission systems. First and second order models are taken into account including the effect of leakage and compressibility on the general system behavior. Chapter three discusses some control techniques using servo and proportional valves. Pump operated and valve operated servo control systems are discussed in this chapter besides to the block diagrams and fluid power symbols. Electro-mechanical controls are depicted in chapter four discussing relays, solenoids and voice coil actuators with their applications in proportional control systems. Chapter five presents the analysis of selected basic hydraulic circuits with corresponding applications while the pneumatic circuits are discussed and analysed in chapter six. The basic principles of hydraulic and pneumatic components are shown in chapter seven. This chapter uses schematic drawings to represent the different designs of pumps, compressors, valves and actuators. Chapter eight tackles a brief overview on some hydraulic and pneumatic applications related to the previously discussed theories. Finally, general maintenance aspects of hydraulic and pneumatic systems are discussed in chapter nine showing the different problems with their possible causes and remedies.

Ahmed Abu Hanieh

Birzeit, August 2010

Contents

Chapter 1

Fundamentals of Fluid Power

1.1 Introduction

Man began to think about different sources of power since the early ages of creation. Fire was the first discovered source of power and energy followed by liquids (namely water) that has been replaced later on by oil. It has been found that gases can have similar behviour under pressure. Liquid and gas phases of material have been given the term *Fluid.* Flowing fluid is considered as a power transmission medium because the continuous flow is capable of changing the existing energy from one shape to another.

Fluid power is a general term used in mechanical engineering for Hydraulics and Pneumatics. This source of energy proved having higher power than any other power handling method which led engineers to consider it as the best solution for heavy duty works. The overall power in fluids depends mainly on the flow and pressure values. The higher the pressure, the smaller the flow to produce the required power for specific task. These pressure and flow levels can be obtained by using pumps in the hydraulic system and compressor in the pneumatic system. The most common known pump is the human heart while the human lungs can be considered as the most famous compressor .

Various industrial applications depend more and more on fluid power, it is even logic to say that fluid power touches every part in engineering applications because of its contribution in automation which is becoming essential to increase productivity of different manufacturing and handling processes. Ease of control can be considered as one of the main advantages of fluid power systems where mechanical, electrical and manual controls are applied easily to the various hydraulic and pneumatic systems. These systems have the ability to amplify the force in high ratio which makes them useful for very high loads and torques. Besides to the fact that they can introduce constant and steady forces and torques leading to high stability and performance. Despite simplicity, safety and high power to weight ratio, fluid power systems suffer from some disadvantages like high noise, liquid leakage and exposure to high pressures that can be dangerous on operators [1].

The first invented manual pump was used to pump water out of wells for the purposes of drinking and irrigation. The basic principle of this pump, shown in Figure 1.1, is based on using a reciprocating positive displacement method where a piston slides inside a cylinder changing the volume occupied by the fluid in every stroke back and forth. A non return valve is used to prevent the water from returning back motion is reversed.

1.2 Power transmission methods

Power transmission systems depend mainly on the source of energy used to produce this power, where they are divided into mechanical , electromechanical , pneumatic and hydraulic . The influence of these power tramission and handling systems is described as follows:

- To obtain high torque to load inertia or power to weight ratio, hydraulics is the best solution followed by pneumatics while mechanical and electromechanical systems give worse results.

- The steady state stiffness of hydraulics is higher than that of the mechanical system while in pneumatics and electromechanical systems stiffness is extremely weak.

Figure 1.1: *Manual pump (Courtesy of King Pumps)*

- On the other hand, the friction level in electromechanical system is much better that all other systems but it is more sensitive to external noise.

The previous discussion shows how much fluid power systems are superior to other power transmission methods but it is worth having a look at a comparison between hydraulics and pneumatics. Table 1.1 shows this comparison [2].

A basic hydraulic system consists mainly of a tank or reservoir for the fluid (oil), a pump , a control valve , an actuator (piston or motor) and a set of pipes as shown in Figure 1.2. The tank in normally open to atmosphere leading to zero gauge pressure of oil at this point. The tank feeds oil to the pump through a filter or strainer. The pump delivers the fluid continuously at a specific flow rate according to its capacity, the accumulation of this flow inside the pipes leads to increase the pressure in the system. A control valve is used here to decide the required operation and direction

Table 1.1: *Comparison between hydraulic and pneumatic systems*

Pneumatics	**Hydraulics**
Fluid is compressible (Air)	Fluid is incompressible (Oil)
Relatively low fluid pressure	Very high fluid pressure
Limited dynamic response	Good dynamic response
Delay time of pistons is big	Very smaller delay time
Higher friction due to dryness	Lower friction due to viscous lubrication
No cavitation effect	Exposed to cavitation
Ability of operation at high temperatures	Temperature is limited to oil characteristics

of fluid. The pressurised fluid is used to extend or retract the piston.

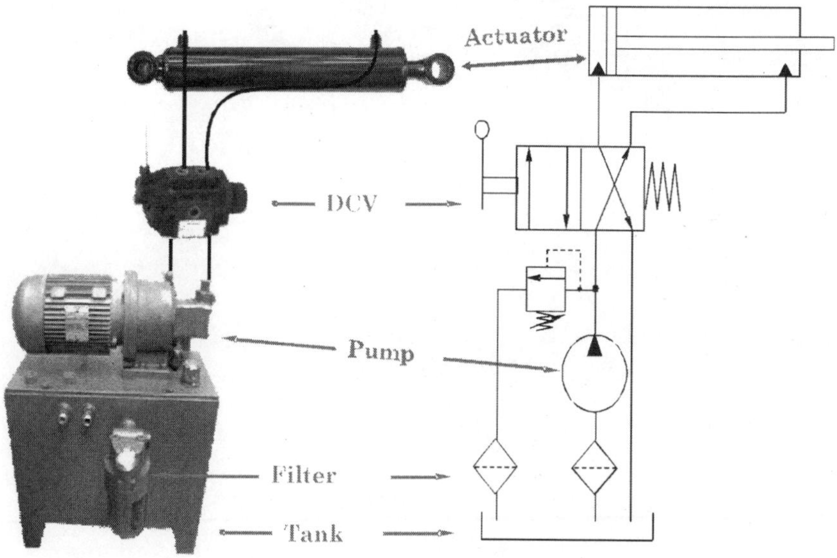

Figure 1.2: *Basic hydraulic system*

A basic pneumatic system includes a compressor , a storage tank for the fluid (air), a control valve, an actuator (piston or motor) and a set of pipes as shown in Figure 1.3. The compressor extracts air from the atmosphere through a filter and compresses it into the storage tank. The pressure inside the tank increases till reaching a specific point where the compressor turns off automatically. The pressurised air flowing in pipes is controlled and directed by means of a control valve that determines the direction and operation of this air that leads, eventually, to extend or retract the actuation piston.

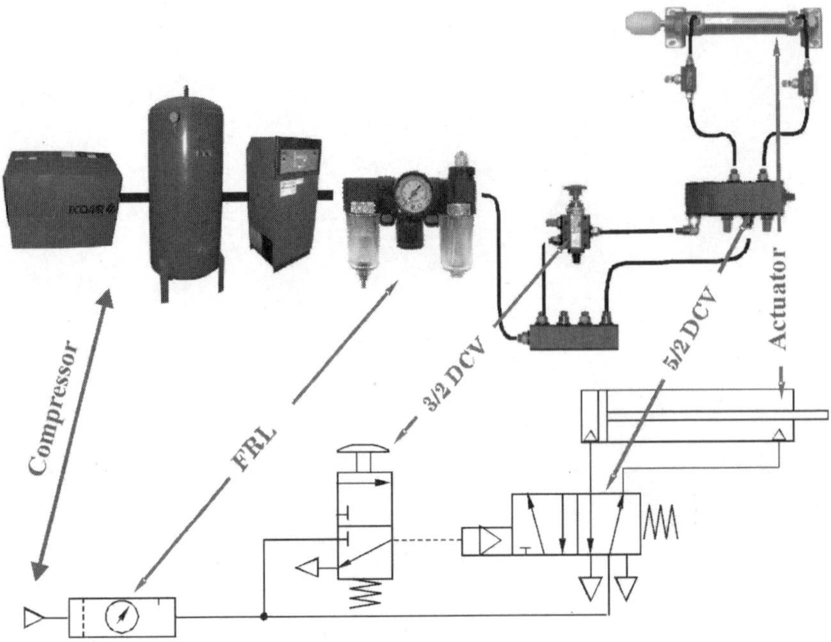

Figure 1.3: *Basic pneumatic system*

1.3 Basic theories

When a system engineer determines to design a fluid power system, he has to take into account the following aspects:

- *Applied mechanics :* Static, kinematic and dynamic calculations are totally involved here.
- *Vibration analysis:* High pressure and temperature accompanied with harmonic disturbances cause a high level of vibrations that can be catastrophic to the mechanical system.
- *Fluid mechanics :* This is considered the main field to be tackled here as far as the systems depend mainly on the motion of fluids.
- *Thermodynamics :* Heat can have a great influence on these systems specially in hydraulics where high temperature reduces, signif-

icantly, the effeciency of hydraulic system by changing oil characteristics.

The fundamental calculations are based on the following basic laws:

- Newton's law.
- Perfect gas law.
- Torricelli's theorem.
- Pascal's law.
- Bernoulli's equation.

1.3.1 Newton's law of motion

Newton's second law of motion states that the induced force F is directly proportional to the acceleration a of a moving material. The proportionality constant is defined as the mass m of the matter [3], [4]:

$$F = ma = m\frac{dv}{dt} \tag{1.1}$$

where v is the velocity of the matter. It is known that the mass is defined by the multiplication between the density ρ and the volume V where, $m = \rho V$. This leads to the equation:

$$F = \rho V\frac{dv}{dt} \quad or \quad F = \rho\frac{dV}{dt}v \tag{1.2}$$

The rate of change of the volume is equal to the fluid flow rate Q which leads to the equation

$$F = \rho Q v \tag{1.3}$$

The foregoing discussion is applied directly in the case of flowing liquid (Hydraulics).

1.3.2 Perfect gas law

In the pneumatic case, the flowing fluid is a gas where there is a high influence of temperature and pressure that cause a change in density and can be governed by the Perfect Gas Law [7]:

$$PV = mRT \tag{1.4}$$

where P is the absolute pressure , T is the temperature in Kelvin and R is the gas constant that can be calculated from the universal gas constant R_u divided by the molecular weight MW of gas:

$$R = \frac{R_u}{MW} \tag{1.5}$$

Knowing that the universal gas constant $R_u = 8315 J/kg.K$ and the molecular weight of air $MW = 28.97$, results in a gas constant value for air equals $R = 287$. Note here that the absolute pressure is calculated by adding the measured gauge pressure to the atmospheric pressure that equals at sea level to 101.3 kPa. and the temperature in Kelvin is calculated by adding the Cellecius temperature to 273.15.

As special cases of the perfect gas law, one can express the following:

- If a given constant mass of a gas is compressed or expanded at a constant temperature, the absolute pressure is inversely proportional to the volume which leads to Boyle's law:

$$P_1 V_1 = P_2 V_2 \tag{1.6}$$

- If a given constant mass of a gas changes its temperature under constant pressure, the volume is directly proportional to the temperature leading to Charle's Law :

$$\frac{V_1}{T_1} = \frac{V_2}{T_2} \tag{1.7}$$

- If a given constant mass of a gas changes its temperature under constant volume, the pressure is directly proportional to the temperature leading to Gay Laussac's Law :

$$\frac{P_1}{T_1} = \frac{P_2}{T_2} \tag{1.8}$$

- Combining the three previous laws in equations (1.6), (1.7) and (1.8), results in the most commonly accepted form called the combined gas law :

$$\frac{P_1 V_1}{T_1} = \frac{P_2 V_2}{T_2} = Constant \tag{1.9}$$

Example 1.1

A pneumatic air reservoir with a capacity of 150 liter is filled with a compressed air at a gauge pressure of 900 kPa at a temperature of 45^oC. The air is cooled to a temperature of 20^oC. Determine the final pressure in the reservoir.

Solution

To calculate the absolute pressure and temperature
$P_1 = 900 + 101.3 = 1001.3$ kPa
$T_1 = 45 + 273.15 = 318.15$ K
$T_2 = 20 + 273.15 = 293.15$ K

The combined gas law in equation (1.9) can be applied here, but the volume is constant, then equation (1.8) can also be applied:

$$P_2 = P_1 \frac{T_2}{T_1} = 1001.3 \frac{293.15}{318.15} = 922.6 \text{ kPa (absolute)}$$

When a system is expanded or compressed at a constant temperature it is said to be $Isothermal$. In the case of accumulator, rapid expansion or compression is done with no heat losses. This process is called *Adiabatic* and can be determined as follows

$$PV^{\gamma} = K = constant \tag{1.10}$$

where

$$\gamma = \frac{C_p}{C_v}$$

knowing that C_p is the specific heat of fluid at constant pressure and C_v is the specific heat of fluid at constant volume. This leads to the following calculation:

$$\left(\frac{T_1}{T_2}\right) = \left(\frac{V_2}{V_1}\right)^{\gamma-1} = \left(\frac{P_1}{P_2}\right)^{\frac{\gamma-1}{\gamma}} \tag{1.11}$$

or

$$\left(\frac{P_1}{P_2}\right) = \left(\frac{V_2}{V_1}\right)^{\gamma} = \left(\frac{T_1}{T_2}\right)^{\frac{\gamma-1}{\gamma}} \tag{1.12}$$

or

$$\left(\frac{V_1}{V_2}\right) = \left(\frac{P_2}{P_1}\right)^{\frac{1}{\gamma}} = \left(\frac{T_2}{T_1}\right)^{\frac{1}{\gamma-1}} \tag{1.13}$$

1.3.3 Torricelli's theorem

Evangelista Torricelli who lived from 1608 to 1647 put forward a theorem stating that; the velocity of a free jet of a fluid is proportional to the square root of the head producing the jet [5]. This theorem is demonstrated in Figure 1.4. Having the same atmospheric pressure S, energy conservation can be applied between points A and B.

$$\frac{1}{2}mv^2 = mgh \tag{1.14}$$

$$v = \sqrt{2gh} \tag{1.15}$$

This leads to the direct relationship between the velocity v and the head h, where the head is related to the liquid pressure P by the relation $h = P/\rho g$.

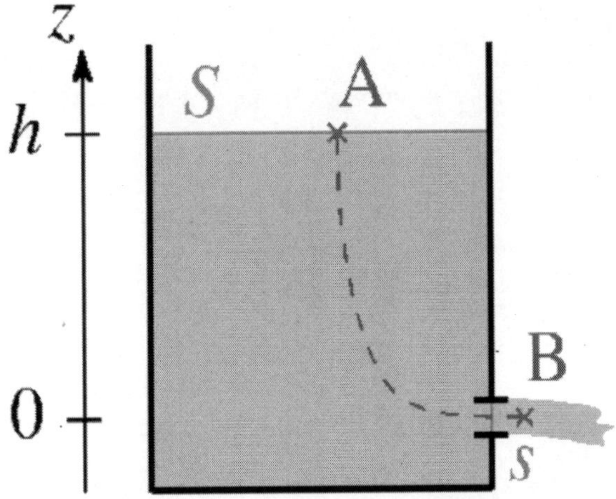

Figure 1.4: *Torricelli's theorem*

1.3.4 Pascal's law

The principle of power transmitted by fluid has been laid by Blaise Pascal, [6]. Power transmission here is similar to power transmission in a human arm muscle or a lever arm as shown in Figure 1.5.

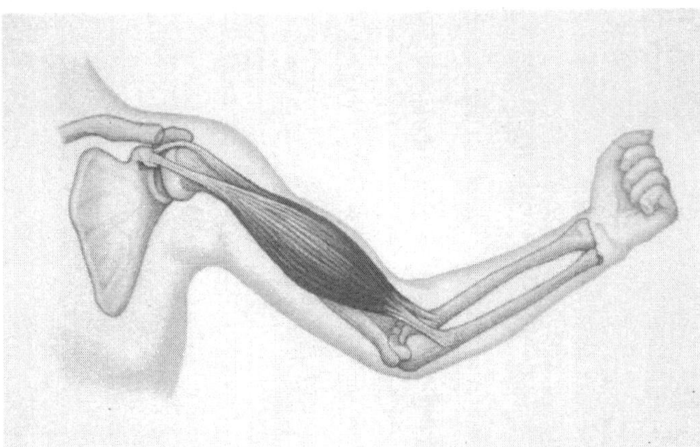

Figure 1.5: *Human arm muscle (Courtesy of Anatomy resources)*

Pascal's theorem states that:

- Fluid pressure has the same value throughout an enclosed fluid in a vessel.
- Pressure acts equally in all directions at the same time.
- Pressure acts at right angle to any surface in contact with the fluid.

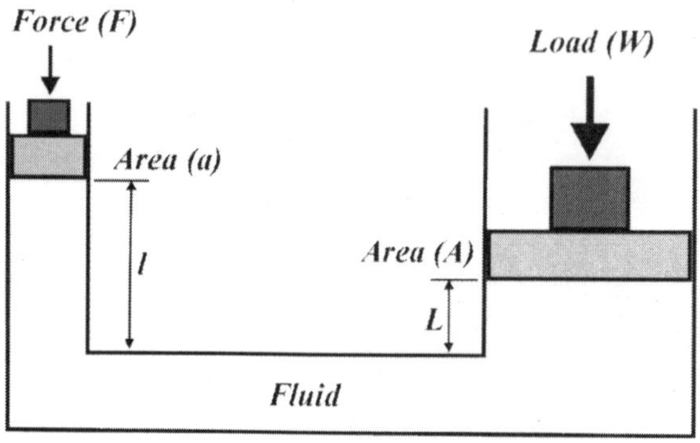

Figure 1.6: *Pascal's Theorem*

The main idea is shown in Figure 1.6. The system consists of a force cylinder with an area a and an acting force F that causes the piston to move a distance l called the pumping ram, it is connected to a load cylinder with an area A and a supported load W that causes this piston to move a distance L called the load ram. The pressure of fluid is calculated as follows:

$$P = \frac{F}{a} = \frac{W}{A} \tag{1.16}$$

or taking the force ratio as

$$\frac{W}{F} = \frac{A}{a} \tag{1.17}$$

Assuming incompressible fluid, the displaced volume V is calculated from $V = AL = al$ and the total work is calculated from

$$Work = PV = PAL \tag{1.18}$$

The SI units for the work are
$Work = P(N/m^2).V(m^3) = N.m$

The mechanical work is estimated by the multiplication of the force and the displacement resulting in an acting force on the left hand side $F = PA$ and a resulting force on the load $W = PA$. It is clear that the magnification of the force depends on the area ratio A/a.
On the other hand, the hydraulic power is equal to $Power = PQ$ where Q is the volume flow per unit time $Q = V/time$. The SI units for the hydraulic power are
$Power = P(N/m^2).Q(m^3/s) = N.m/s = Watt$

In practical application, this theorem is applied to the fluid power cylinder (linear actuator) where the commercially used units are based on expressing the flow rate in liter per minutes (l/min) and pressure in (bar)

$$Q(l/min) = \frac{Q}{60}(l/s) = \frac{Q}{60 \times 10^3}(m^3/s)$$

$$P(bar) = P \times 10^5 (N/m^2)$$

The hydraulic power

$$Power = Q(l/min)(\frac{1}{60 \times 10^3})(m^3/s) \times P(bar)(1 \times 10^5)(N/m^2)$$

$$= \frac{QP}{600}10^3(N.m/s) = \frac{QP}{600}10^3(Watts)$$

$$= \frac{Q(l/min) \times P(bar)}{600} = Power(kW)$$

Example 1.2

Consider the hydraulic piston shown in Figure 1.7 used to push a load of 1000 N. The piston is actuated by a manual pump that compresses a fluid to the piston through a pipe with inside diameter of 10 mm. The diameter of the cylinder is 100 mm and the diameter of the rod is 40 mm. Determine the human force needed to act on the piston
a- in Extension stroke.
b- in Retraction stroke.

Solution

The pressure is assumed to have the same value in pipe and cylinder. The area of the pipe is

$$A_{pipe} = \frac{\pi}{4}(0.01)^2 = 0.0000785m^2$$

a- For the extension stroke, the area of the piston side A_p is

$$A_p = \frac{\pi}{4}(0.1)^2 = 0.00785m^2$$

Neglecting the lever arm effect, the hand force F_{hand} needed to act on the manual handle in this case is

$$F_{hand} = \frac{A_{pipe}}{A_p}(Load) = \frac{0.0000785}{0.00785}(1000) = 10N$$

b- For the retraction stroke, the area of the rod side A_r is

$$A_r = \frac{\pi}{4}[(0.1)^2 - (0.04)^2] = 0.006594m^2$$

Neglecting the lever arm effect, the hand force F_{hand} needed to act on the manual handle in this case is

$$F_{hand} = \frac{A_{pipe}}{A_r}(Load) = \frac{0.0000785}{0.006594}(1000) = 11.9N$$

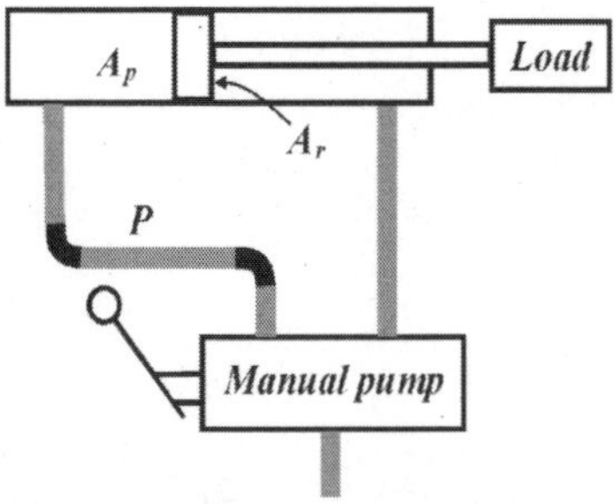

Figure 1.7: *A hydraulic piston, for example 1.2*

1.3.5 Bernoulli's equation of energy

Daniel Bernoulli who lived between 1700 to 1782 developed the concept of energy conservation for fluids flowing in a pipe with changing cross-sectional area and connecting between two different levels as shown in Figure 1.8. Conservation of energy concept means that energy can neither be created nor destroyed. When a pipe is installed between different elevations, the total energy of fluid (kinetic , potential , hydraulic and losses) is conserved throughout the pipe [5].

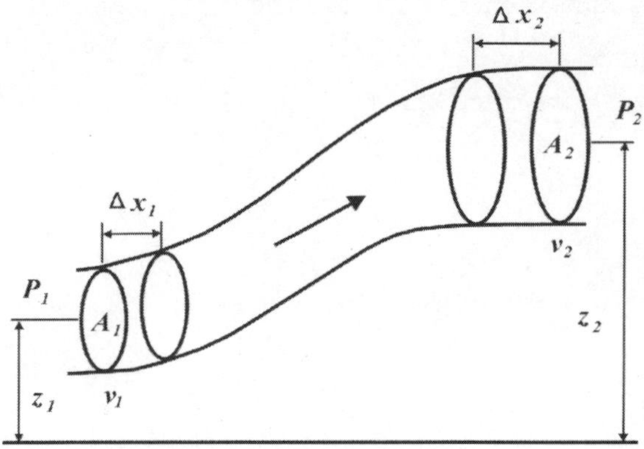

Figure 1.8: *Bernoulli's Theorem*

Consider an incompressible, nonviscous, pressurized fluid flowing in a pipe from point (1) at an elevation z_1 with a cross-sectional area A_1 and a pressure P_1 to point (2) at an elevation z_2 with a cross-sectional area A_2 and a pressure P_2. The force at the input of the pipe is the multiplication of P_1 and A_1 and the volume of fluid pushed to the pipe a displacement Δx_1 is the multiplication of A_1 and Δx_1 [6]:

$$\Delta V_1 = A_1 \Delta x_1$$

Similarly, the volume of fluid reaching at the exit of the pipe is

$$\Delta V_2 = A_2 \Delta x_2$$

The incompressibility of the fluid means that the volume remains the same throughout,

$$\Delta V_1 = \Delta V_2 = \Delta V$$

Hence, calculating the work done at the inlet and outlet of the pipe

$$Work_{in} = P_1 A_1 \Delta x_1 = P_1 \Delta V$$

$$Work_{out} = P_2 A_2 \Delta x_2 = P_2 \Delta V$$

The change in kinetic energy between the inlet and outlet is given by

$$\Delta KE = \frac{1}{2}\Delta m_2 v_2^2 - \frac{1}{2}\Delta m_1 v_1^2 \tag{1.19}$$

where the mass at the inlet is

$$\Delta m_1 = \rho A_1 \Delta x_1$$

Similarly, the mass of fluid reaching at the exit of the pipe is

$$\Delta m_2 = \rho A_2 \Delta x_2$$

The conservation of mass (no leakage) leads to the fact that the mass remains the same throughout the pipe,

$$\Delta m_1 = \Delta m_2 = \Delta m$$

The change of potential energy due to weight as a function of the height measured from an inertial reference reads

$$\Delta PE = \Delta m_2 g z_2 - \Delta m_1 g z_1 = \Delta m g(z_2 - z_1) \tag{1.20}$$

Applying the rule of conservation of energy

$$\text{Work in} = \text{Kinetic Energy} + \text{Potential Energy} + \text{Work out}$$

$$P_1 \Delta V = \Delta KE + \Delta PE + P_2 \Delta V$$

$$P_1 \Delta V = \frac{1}{2}\Delta m(v_2^2 - v_1^2) + \Delta m g(z_2 - z_1) + P_2 \Delta V \tag{1.21}$$

But it is known that $\Delta m/\Delta V = \rho$, thus, dividing equation (1.21) by ΔV results in

$$P_1 = P_2 + \frac{1}{2}\rho v_2^2 + \rho g(z_2 - \frac{1}{2}v_1^2) - \rho g z_1 \tag{1.22}$$

On the other hand, dividing equation (1.21) by Δt, knowing that $\Delta V/\Delta t = Q$ and adding the effect of friction losses leads to the equation

$$P_1 Q = \frac{1}{2}\rho Q(v_2^2 - v_1^2) + \rho g Q(z_2 - z_1) + P_2 Q + \textit{friction losses} \tag{1.23}$$

Eventually, equation (1.22) can be divided by the quantity ρg that defines the specific weight of the fluid γ. This leads to the general form of Bernoulli's equation. For hydraulic applications, the pump head H_p, motor head H_m and friction losses head H_L are added to the equation to form the general modified form of Bernoulli's equation

$$z_1 + \frac{P_1}{\gamma} + \frac{v_1^2}{2g} + H_p - H_m - H_L = z_2 + \frac{P_2}{\gamma} + \frac{v_2^2}{2g} \tag{1.24}$$

Example 1.3

Consider the hydraulic system shown in Figure 1.9, the system consists of a hydraulic pump fed by an oil tank through a filter or strainer. The pump delivers a flow of 120 l/min adding a power of 3.7 kW. The pipe has a 25 mm inside diameter and transports oil with specific gravity 0.9. The oil moves a hydraulic motor and returns back to the tank. considering the portion of the system between point (1) on the oil surface in the tank and point (2) at a height of 6 m on the inlet of the hydraulic motor, dertermine the pressure at point (2) knowing that the head loss H_L between points 1 and 2 is 9 m of oil.

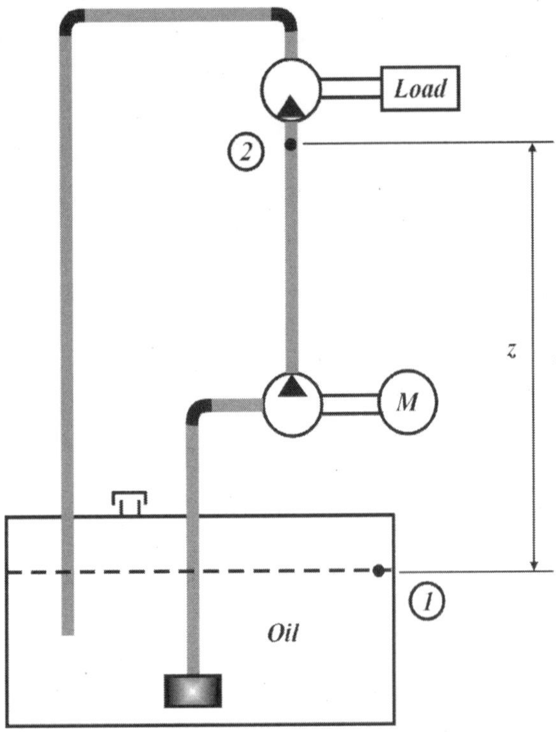

Figure 1.9: *Hydraulic system for example 1.3*

Solution

Applying Bernoulli's equation of conservation of energy

$$z_1 + \frac{P_1}{\gamma} + \frac{v_1^2}{2g} + H_p - H_m - H_L = z_2 + \frac{P_2}{\gamma} + \frac{v_2^2}{2g}$$

Looking at the different terms of the equation

- The hydraulic motor is not included in the calculations, so $H_m = 0$.
- The oil surface is considered large enough and its motion is negligible, so $v_1 = 0$.
- The oil tank is open to the atmosphere, thus, gauge pressure $P_1 = 0$.
- The reference of height is taken at point 1, so the head of point 1 $z_1 = 0$.
- The head of point 2 is given as $z_2 = 6m$.
- The head of friction losses in pipes is given by $H_L = 9m$.

Substituting the previous values in Bernoulli's equation

$$0 + 0 + 0 + H_p - 0 - 9 = 6 + \frac{P_2}{\gamma} + \frac{v_2^2}{2g}$$

This leads to

$$\frac{P_2}{\gamma} = H_p - \frac{v_2^2}{2g} - 15$$

Calculating γ

$$\gamma = S.G\gamma_{water}$$

$$\gamma = (0.9)(9800) = 8820\ kg/m^2s^2$$

The flow rate

$$Q = 120(l/min)(min/60s)(m^3/1000l) = 0.002\ m^3/s$$

The head of the pump

$$H_p = \frac{Power}{\gamma Q}$$

$$H_p = \frac{3.7 \times 10^3}{(8820)(0.002)} = 209.7\ m$$

The inside area of the pipe

$$A = \frac{\pi}{4} D^2 = \frac{\pi}{4}(0.025)^2 = 0.00049\ m^2$$

The velocity at point 2

$$v_2 = \frac{Q}{A} = \frac{0.002}{0.00049} = 4.1\ m/s$$

The velocity effect on Bernoulli's equation

$$\frac{v_2^2}{2g} = \frac{(4.1)^2}{2 \times 9.8} = 0.85\ m$$

Substituting again in the equation of energy

$$\frac{P_2}{\gamma} = 209.7 - 0.85 - 15 = 193.85\ m$$

Finally, solving for the pressure P_2

$$P_2 = (193.85)(8820) = 1.7 \times 10^6\ Pa = 1.7\ MPa$$

1.4 Friction losses

1.4.1 Friction losses in hydraulic systems

Figure 1.10 shows a general hydraulic system consists mainly of a hydraulic pump driven by an electric motor and a douple acting hydraulic piston. The pump's suction line takes the oil from the reservoir after being filtered by a filter or strainer. A check (non return) valve is installed between the pump and the tank to prevent oil from returning back to the tank in

stall conditions. The direction and speed of fluid flow is controlled by the directional control valve. This operation involves a mechanical energy coming in to the system from the electric motor, a mechanical energy going out of the system by the hydraulic piston and a heat energy dissipated from the system due to friction losses. As reducing heat losses needs expensive changes in different parts, designers try to find a compromise between loss reduction and high costs [1].
The dissipated heat caused by friction between fluid and pipe depends on

- The roughness of the path: the more tortous the path, the greater the losses.
- The pipe dimensions: the smaller the pipe diameter and the longer the pipe, the greater the losses.
- The viscosity of the fluid: the higher the fluid viscosity, the greater the losses

Reynolds number

When a fluid passes through a pipe, the velocity of fluid layer near the walls of the cylinder reaches zero while the maximum velocity occurs at the greatest distance from the wall which is at the centerline of the pipe. There are three shapes of fluid flow; Laminar, transient and turbulent. The determination of the type of flow depends mainly on the fluid velocity and viscosity from one side and the size and shape of the pipe from the other side. The most common indicator used to determine the type of flow is *Reynolds number Re*. Osborn Reynolds in 1833 carried out a set of experiments in which he found that the nature of flow depends on a dimensionless number that he called *Reynolds number Re*. The value of Reynolds number in the transition form between laminar and turbulent varies as follows:

$$2000 \leq Re \leq 4000 \textit{ for smooth surface (new pipe)}$$

$$1200 \leq Re \leq 2500 \textit{ for corrugated surface (old pipe)}$$

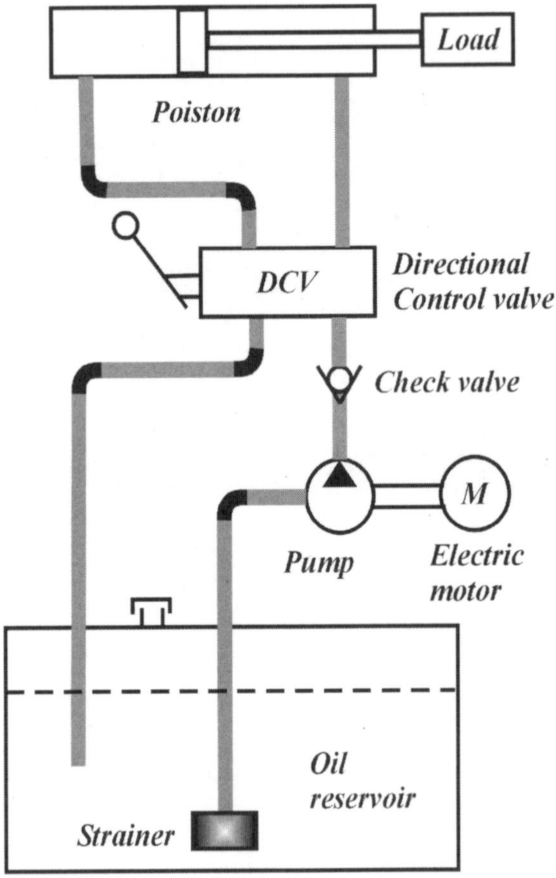

Figure 1.10: *General hydraulic system with linear actuator*

Reynolds number can be calculated from the following equation:

$$Re = \frac{vD}{\nu} = \frac{vD\rho}{\mu} \tag{1.25}$$

where v is the fluid velocity, ρ is the fluid density, D is the hydraulic diameter of the pipe, ν is the kinematic viscosity and μ is the dynamic viscosity.

The pipe's cross sectional shape is not always circular, it can have any shape. Thus, the hydraulic diameter is calculated by

$$D = \frac{4 \times flow\ section\ area}{flow\ section\ perimeter} \tag{1.26}$$

The viscosity of a fluid is the measure of its resistance to flow. High viscosity of fluid leads to energy dissipation during fluid flow. Dynamic viscosity μ is the ratio between the shear stress and the shear rate. Kinematic viscosity ν can be calculated from the dynamic viscosiy

$$\nu = \frac{\mu}{\rho} \tag{1.27}$$

The dimension of ν is given in Stoke St or m^2/s, where $1St = cm^2/s = 10^{-4}m^2/s$. In smaller units, $1cSt = 10^{-6}m^2/s$. Similarly, dimension of μ is given by $Poise$, where $1kg/m.s = 10Poise$. In smaller units, $1Poise = 1g/cm.s$.

Example 1.4

Consider an oil flowing at a rate of 1.6 l/s in an equilateral triangular pipe with a side $y = 15mm$. If the kinematic viscosity of the oil is 35 cSt, find Reynolds number and discuss the results.

Solution

$$Cross\ sectional\ area = \frac{y\sqrt{y^2 - (y/2)^2}}{2} = 9.743 \times 10^{-5}m^2$$

$$Perimeter = 3y = 0.045m$$

$$D = \frac{4 \times 9.743 \times 10^{-5}}{0.045} = 0.00866m$$

$$v = \frac{Q}{A} = \frac{1.6 \times 10^{-3}}{9.743 \times 10^{-5}} = 164.2m/s$$

$$Re = \frac{164.2 \times 0.00866}{35 \times 10^{-6}} = 4063.3$$

The value of Reynolds numbers shows that the flow is turbulent which implies the necessity to increase the pipe diameter in order to have a laminar flow to avoid pressure drop and high noise and to reduce energy dissipation, although some designers donot desire this solution.

Darcy's equation

The main cause of head loss in fluid power systems is friction. The friction losses are dissipated to the environment in a shape of heat energy due to direct contact between fluid and pipes or fittings . these losses are transformed into head H_L that can be calculated in pipes, for both laminar and turbulent flow, from Darcy's equation

$$H_L = f(\frac{L}{D})(\frac{v^2}{2g}) \tag{1.28}$$

where
f = friction factor (dimensionless)
L = length of pipe (m, ft)
D = inside diameter of pipe (m, ft)
v = fluid velocity (m/s, ft/s)
g = gravitational acceleration (m/s^2, ft/s^2)

Friction factor f for laminar flow can be calculated by

$$f = \frac{64}{Re} \tag{1.29}$$

This results in calculating the head loss for laminar flow as follows

$$H_L = \frac{64}{Re}(\frac{L}{D})(\frac{v^2}{2g}) \tag{1.30}$$

friction factor f for turbulent flow can be taken from Moody diagram shown in Figure 1.11.

Moody diagram shows that the friction factor f can be picked out from one of the curves depending on Reynold's number from one side and the relative roughness from another side. Relative roughness can be estimated by

$$Relative\ Roughness = \frac{\epsilon}{D} \tag{1.31}$$

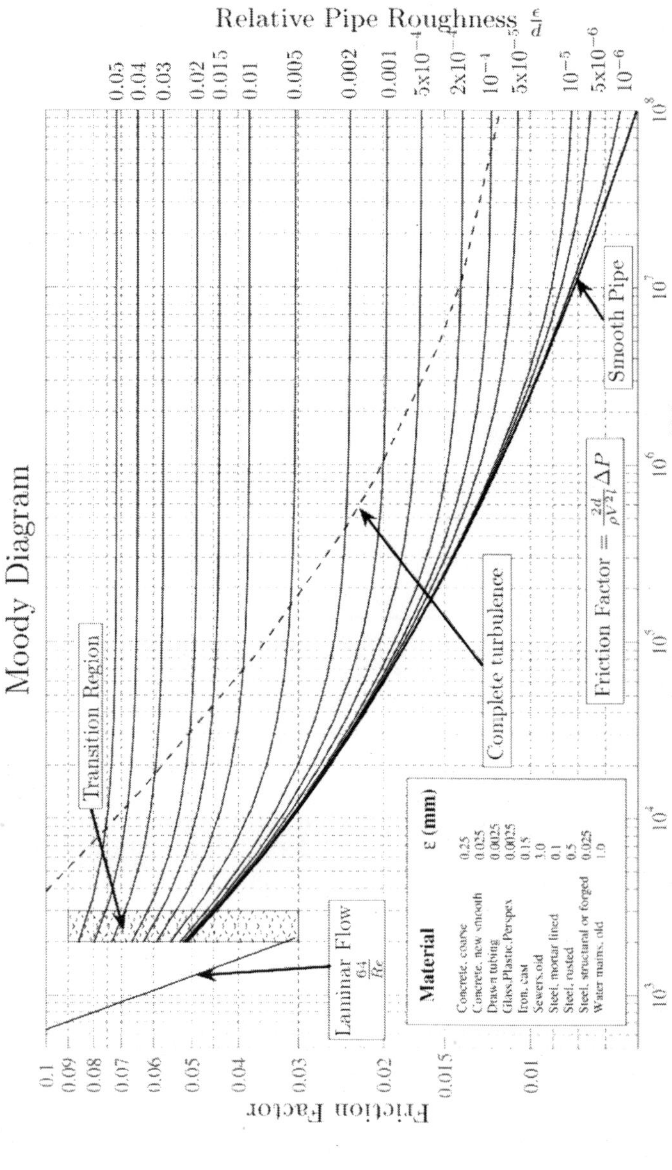

Figure 1.11: *Moody Diagram (courtesy of University of Sheffield)*

where D is the inside diameter of the pipe and ϵ is the height of the insdie corrugation in the pipe. ϵ values for different materials are given in the table included in Figure 1.11. Moody diagram has the following charcteristics:

- Moody diagram is plotted on a log-log scale because of high differences in values.
- The transient flow where Reynold's number is between 2000 and 4000 has no clear values because flow is not possibly predicted.
- At low Reynold's number values (less than 2000), flow is laminar, thus, friction coefficient is a constant value $f = 64/Re$.
- At high Reynold's number values (more than 4000), flow is turbulent, thus, friction coefficient is picked out of the curve where the values of Re and ϵ/D intersect. Interpolation is needed for further accuracy.

Valves and Fittings

It is almost impossible to have a fluid power circuit without valves or fittings. Elbows , tees and return bends are the most common fittings needed to branch or direct a pipeline. On the other hand, globe valves, gate valves, check valves and directional control valves have high influence on the head losses in the system. To include the effect of these valves and fittings in the head loss of the system, equation (1.28) has been truncated into the following equation

$$H_L = K\frac{v^2}{2g} \tag{1.32}$$

where K is the constant of proportionality or K factor. It is clear from the relationship between equation (1.32) and equation (1.28) that the ratio L/D has been set to a unit and the main factor influencing here is K. The proportionality constant K has been determined experimentally and is given in Table 1.2 for the different fittings shown in Figure 1.12.

Table 1.2: *K factor for different fittings and valves*

Fitting	**K factor**
Globe valve	
Wide open	10.0
Half open	12.5
Gate valve	
Wide open	0.19
3/4 open	0.9
1/2 open	4.5
1/4 open	24
Return bend	2.2
Standard tee	1.8
Standard elbow	0.9
45° elbow	0.42
90° elbow	0.75
Check valve	4.0

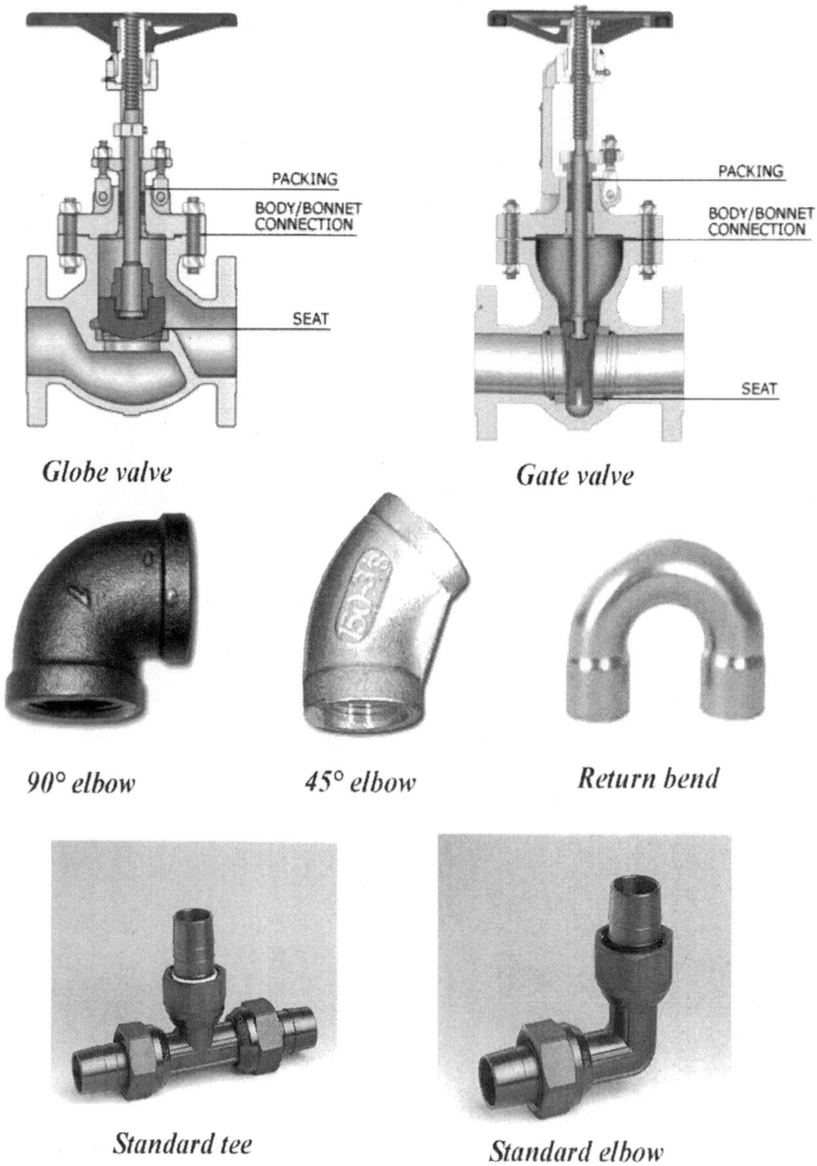

Figure 1.12: *Fittings and valves (Courtesy of PAR and Y & T Metals)*

In the case of specific job valves like directional control valves and different types of spool valves, empirical curves are plotted for each valve by the manufacturer depending on the design and characteristics of the valve itself. Figure 1.13 shows a cutaway of a spool directional control valve. These curves give the relationship between the flow rate passing through the valve and the pressure drop across the valve due to friction and flow resistance.

Figure 1.13: *Cutaway of directional control valve (Courtesy of Hoyea Inc.)*

Equivalent length

Valves and fittings are usually istalled in a fluid power circuit and connected to pipes with specific diameter D. An easy way to estimate the head loss of valves and fittings is to assume replacing them with a portion of pipe having the same diameter D and an equivalent length L_e. This can be done by assuming the same losses in fittings and equivalent pipe

$$H_{L(fitting)} = H_{L(pipe)} \tag{1.33}$$

Substituting the corresponding values for both sides from equation (1.28) and equation (1.32) gives

$$K(\frac{v^2}{2g}) = f(\frac{L_e}{D})(\frac{v^2}{2g}) \tag{1.34}$$

Eliminating velocities from both side, since they have the same value results in

$$L_e = \frac{KD}{f} \tag{1.35}$$

The value of equivalent length in equation (1.35) can be used then in equation (1.28) to calculate the head loss in valve or fitting or, one can add the value of L_e to the total length of the pipe before applying equation (1.28).

Example 1.5

Consider the hydraulic system in Figure 1.14. The pump adds a power of 3.73 kW to a fluid of specific gravity equals to 0.9 and a kinematic viscosity of 100 cSt. The fluid flow rate is 0.00190 m^3/s. The pipe has a 25.4 mm inside diameter and dimensions of $L_1 = 1.22m$, $L_2 = 0.3m$ and $L_3 = 4.88m$. Knowing that the pressure ar point (1) is atmospheric pressure and taking into account the friction losses, determine the pressure at point (2).

Solution

Applying Bernoulli's equation of motion for the conservation of energy

$$z_1 + \frac{P_1}{\gamma} + \frac{v_1^2}{2g} + H_p - H_m - H_L = z_2 + \frac{P_2}{\gamma} + \frac{v_2^2}{2g}$$

Finding the different parameters of Bernoulli's equation
$z_1 = 0$ (*Height reference*)
$H_m = 0$ (*Hydraulic motor is not included*)
$v_1 = 0$ (*Large oil reservoir*)

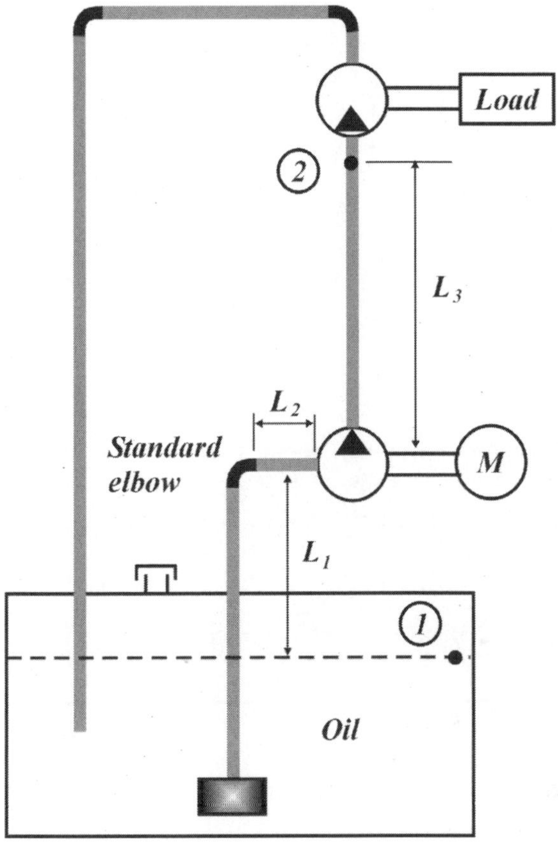

Figure 1.14: *Hydraulic system with friction effect, for example 1.5*

$P_1 = 0$ (*Oil tank is open to atmosphere*)

$$z_2 = L_1 + L_3 = 1.22 + 4.88 = 6.1m$$

$$v_2 = \frac{Q}{A} = \frac{0.0019}{(\pi/4)(0.0254)} = 3.74m/s$$

The contribution of velocity to the head at point (2)

$$\frac{v_2^2}{2g} = \frac{3.74}{2(9.8)} = 0.714m$$

The head of the pump is calculated by

$$H_p = \frac{Pump\ hydraulic\ power}{\gamma \times Q}$$

$$H_p = \frac{3730}{(0.9)(9800) \times (0.0019)} = 223.1m$$

To determine the head loss due to friction, Reynold's number must be calculated at point (2)

$$Re = \frac{vD}{\nu} = \frac{(3.74)(0.0254)}{100 \times 10^{-6}} = 944$$

The flow at point (2) is laminar and the friction coefficient f is given by

$$f = \frac{64}{Re} = \frac{64}{944} = 0.0678$$

Calculating the equivalent length for the standard elbow (K=0.9)

$$L_e = \frac{KD}{f} = \frac{(0.9)(0.0254)}{0.0678} = 0.337m$$

The total length is

$$L_T = L_1 + L_2 + L_3 + L_e = 1.22 + 0.3 + 4.88 + .337 = 6.737m$$

The head loss is

$$H_L = f(\frac{L_T}{D})(\frac{v^2}{2g}) = (0.0678)(\frac{6.737}{0.0254})(0.714) = 12.9m$$

Substituting in Bernoulli's equation

$$0 + 0 + 0 + 223.1 - 0 - 12.9 = 6.1 + \frac{P_2}{\gamma} + 0.714$$

$$\frac{P_2}{\gamma} = 203.4m$$

Solving for P_2

$$P_2 = \gamma \times 203.4 = (0.9)(9800)(203.4) = 1790000Pa = 1.79MPa$$

1.4.2 Friction losses in pneumatic systems

Pressure losses in pneumatic systems can be calculated from Harris formula. Harris formula is an empirical formula used to calculate pressure loss directly in British units as follows:

$$P_f = \frac{CLQ^2}{3600(CR)D^5} \tag{1.36}$$

where
P_f = Pressure loss (psi).
L = Length of pipe (ft).
Q = flow rate (ft^3/min).
D = Inside diameter of pipe (in).
CR = Compression ratio,

$$CR = \frac{Pressure\ in\ pipe}{Atmospheric\ pressure}$$

C = Experimentally determined coefficient, which is for schedule 40 commercial pipes

$$C = \frac{0.1025}{D^{0.31}}$$

Substituting this value in Harris formula, equation (1.36) results in

$$P_f = \frac{(0.1025)LQ^2}{3600(CR)D^{5.31}} \tag{1.37}$$

This formula can be used directly to calculate the pressure loss in pneumatic pipes.

1.5 References

[1] A. Esposito, *Fluid Power with Applications*, Prentice Hall, 2003.

[2] A. Parr, *Hydraulics and pneumatics*, Newnes, 1991.

[3] R. Hebbeler, *Engineering Mechanics-Dynamics*, Prentice Hall, 1998.

[4] F. Beer and E. Johnston, *Vector Mechanics for Engineers,Dynamics*, SI Metric Edition, McGraw-Hill, 1977.

[5] I. Shames, *Mechanics of Fluids*, Second Edition, McGraw-Hill, 1982.

[6] J. Francis, *Fluid Mechanics for Engineering Students*, Fourth Edition, Edward Arnold, 1975.

[7] G. Van Wylen and R. Sonntag, *Fundamentals of Classical Thermodynamics*, Third Edition, John Wiley and Sons, 1985.

Chapter 2

Modelling of Fluid Power Systems

2.1 Introduction

To study any mechanical system, it is important to represent this system using mathematical equations. The most modern way of representation is expressed in building a second order differential equation of motion for the system which is called *Modelling* [1] . Despite the fact that most of the fluid power systems have much higher order equations. This equation of motion can be built using one of the following ways:

- Newton's second law of motion which states that the induced force is directly proportional to the acceleration of the body and the summation of external forces acting on the body is equal to the inertial force of the body itself.

- Conservation of energy : where the energy of the body is conserved but it changes from one shape to another as the body changes its position or state.

- Finite element modelling is the most precise numerical way of modelling. It is handled by dividing the body into small pieces to be analysed and studied. this can represent the virtual real behaviour of the system.

- Modal analysis is performed by installing measurement sensors on different parts of the system. The signals retrieved from these sensors are inserted into a mathematical calculation to simulate the real behaviour of the system.

This book will focus on using Newton's law of motion to establish equations of motion .

2.2 Block diagrams

Solving the equation of motion leads to build relations between the different parameters of the systems. These relations are called the "Transfer Functions" . Building an equation of motion and deriving transfer function will be discussed later in this chapter. Transfer function expresses the relation between an input and an output of the system taking into account the influence of all other parameters. Transfer function is usually denoted by the letter G or other letters. It is a time dependent relationship and expressed in Laplace transfrom to be used in frequency domain calculations $G(s)$. The relationships and operations between transfer functions can be studied by using block diagrams [2]. Figure 2.1 shows a block diagram of three transfer functions in series where they can be represented by the multiplication of the functions. Transfer functions in parallel are represented by the summation of these functions as shown in Figure 2.2. In the case of parallel connection, the point where the inputs branch before entering to the blocks is called *Take-off* point and the point of summing the outputs is called *Sum* point.

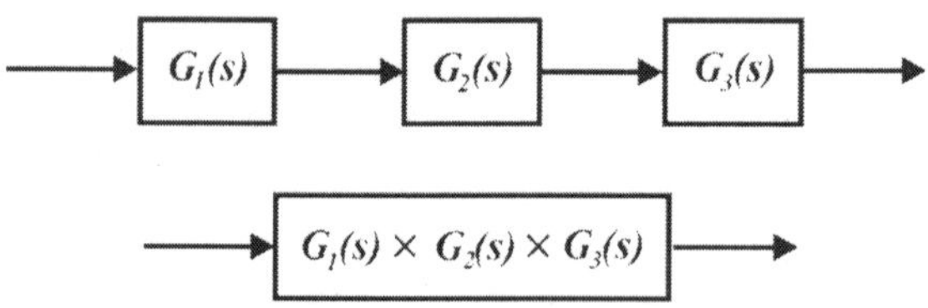

Figure 2.1: *Blocks in series*

The aim of the control system is to minimize the error by seeking for zero value between the input and the signal fed back from the output. Such a system is called *Feedback system* and is shown in Figure 2.3. The feedback control system shown in Figure 2.3 consists of two transfer functions $G_1(s)$

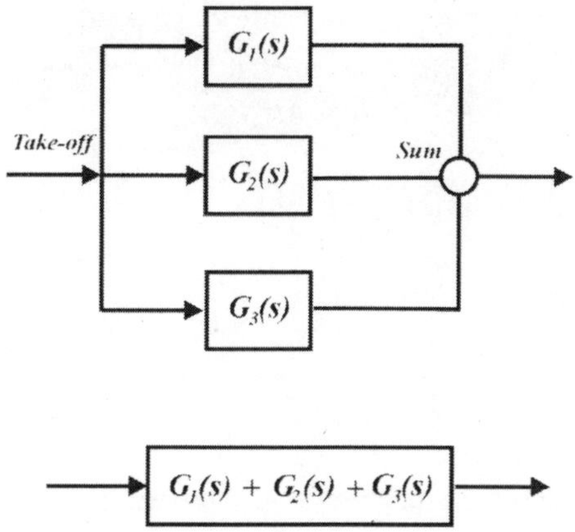

Figure 2.2: *Blocks in parallel*

and $G_2(s)$ connected in series. The input reference signal coming into the system is $I(s)$ and the output signal is $O(s)$. The error signal $E(s)$ is minimized by feeding the output signal back and subtracting it from the input after multiplying by the compensator transfer function $H(s)$ yielding the feedback signal $F(s)$. The transfer function of the system is the ratio between the input $I(s)$ and the output $O(s)$ can be found by manipulating the blocks to find $(O(s)/I(s))$.

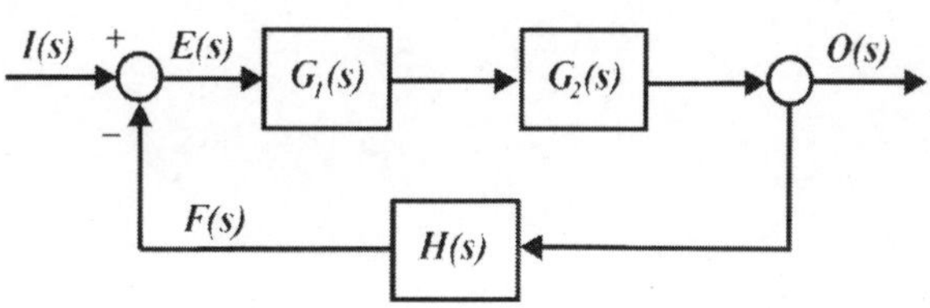

Figure 2.3: *Feedback control system*

Looking at the main feedforward path,

$$O(s) = G_1(s)G_2(s)E(s)$$

Looking at the feedback path,

$$F(s) = H(s)O(s)$$

The error signal at the summation junction,

$$E(s) = I(s) - F(s)$$

Hence,

$$E(s) = I(s) - H(s)O(s)$$

and

$$O(s) = G_1(s)G_2(s)[I(s) - H(s)O(s)]$$

This leads to the ratio between $I(s)$ and $O(s)$

$$\frac{O(s)}{I(s)} = G_1(s)G_2(s) - G_1(s)G_2(s)\frac{H(s)O(s)}{I(s)}$$

Or,

$$\frac{O(s)}{I(s)}[1 + G_1(s)G_2(s)] = G_1(s)G_2(s)$$

Finally,

$$\frac{O(s)}{I(s)} = \frac{G_1(s)G_2(s)}{1 + G_1(s)G_2(s)H(s)} \tag{2.1}$$

This transfer function is considered the general closed-loop transfer function for the feedback control system [3].

Example 2.1

Find the transfer function between u as an input and y as an output in the block diagram shown in Figure 2.4.

Solution
Starting at the summation junction

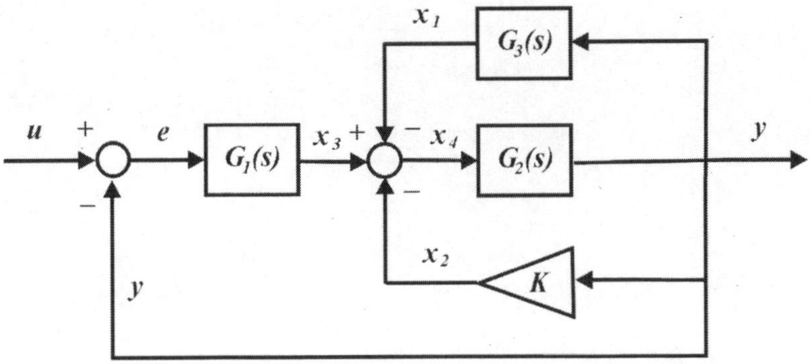

Figure 2.4: *Block diagram for example 2.1*

$$x_4 = x_3 - x_2 - x_1$$

$$x_4 = x_3 - G_3(s)y - Ky$$

But,

$$y = G_2(s)x_4 = G_2(s)[x_3 - G_3y - Ky]$$

Manipulation gives

$$y = \frac{G_2(s)}{1 + G_2(s)[G_3(s) + K]}x_3$$

Looking at the other summation junction

$$e = u - y$$

and

$$x_3 = G_1(s)e$$

then,

$$y = \left(\frac{G_1(s)G_2(s)}{1 + G_2(s)(G_3(s) + K)}\right)e$$

Or,

$$y = \left(\frac{G_1(s)G_2(s)}{1 + G_2(s)(G_3(s) + K)} \right) (u - y)$$

Thus,

$$y = u \left(\frac{G_1(s)G_2(s)}{1 + G_2(s)(G_3(s) + K)} \right) / \left(1 + \frac{G_1(s)G_2(s)}{1 + G_2(s)(G_3(s) + K)} \right)$$

Final mnipulation gives the transfer function

$$\frac{y}{u} = \frac{G_1(s)G_2(s)}{1 + G_2(s)[G_3(s) + G_1(s) + K]}$$

2.3 Conceptual modelling of a transfer function

As mentioned at the beginning of this chapter, Newton's law will be used to build the transfer function. Assuming a lumped parameter model, mass, spring and damper are the basic elements of this model [4]. The Inertia force is given by

$$F = m\frac{d^2x}{dt^2} = m\frac{dv}{dt} \tag{2.2}$$

where, F is the force, m is the mass and x is the displacement, reciprocating the equation gives

$$v = \frac{1}{m}\int F dt \tag{2.3}$$

where v is the velocity. In the case of rotation, the mass is replaced by the Inertia I , the force is replaced by the torque T and the displacement is replaced by the angle θ

$$T = I\frac{d^2\theta}{dt^2} = I\frac{d\omega}{dt} \tag{2.4}$$

Where ω is the angular velocity. The mass or inertia element is a kinetic energy storage element. The spring is considered as a potential energy storage element, the spring force reads

$$F = k(x_1 - x_2) \tag{2.5}$$

or,

$$F = k \int (v_1 - v_2) dt \tag{2.6}$$

In the rotation case

$$T = k(\theta_1 - \theta_2) \tag{2.7}$$

When a damper is added to the system, it bahaves as an energy dissipation element in the form of heat, the damping force is given by

$$F = C(v_1 - v_2) \tag{2.8}$$

In the rotation case

$$T = C(\omega_1 - \omega_2) \tag{2.9}$$

2.4 Effort and flow

The modelling of inputs and outputs of any system can be represented by effort e and flow f as shown in Figure 2.5. The overall power is the multiplication of the effort and the flow

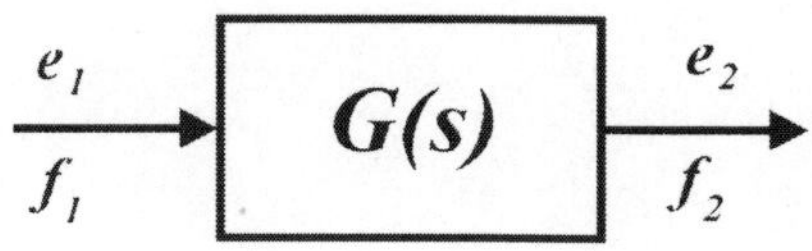

Figure 2.5: *Effort and flow representation*

$$Power = effort \times flow = e \times f$$

and the efficiency is

$$Efficiency = \frac{Power_{out}}{Power_{in}}$$

The ratio between the effect and the flow is called the impedance where

$$Input\ Impedance = \frac{e_1}{f_1}$$

$$Output\ Impedance = \frac{e_2}{f_2}$$

The subscript 1 stands for the input and 2 stands for the output. If the flow is constant $f_1 = f_2 = f$, then

$$Transfer\ Function\ \frac{e_2}{e_1} = \frac{e_2}{f} \times \frac{f}{e_1}$$

According to causality theorem, an effort F can be a cause that leads to effect v (Figure 2.6) and a flow v can be a cause leading to effect F (Figure 2.7).

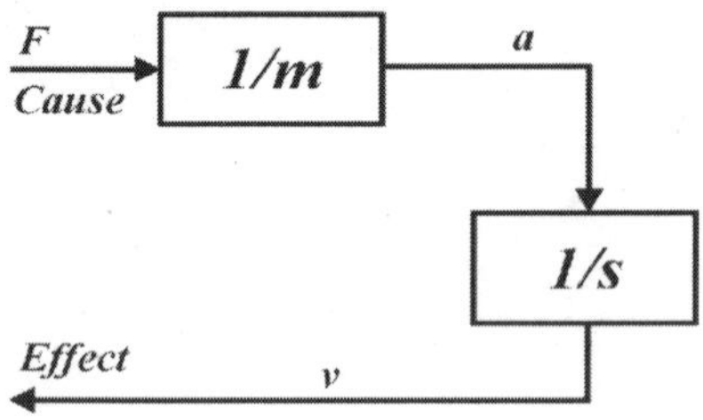

Figure 2.6: *Effort causes flow*

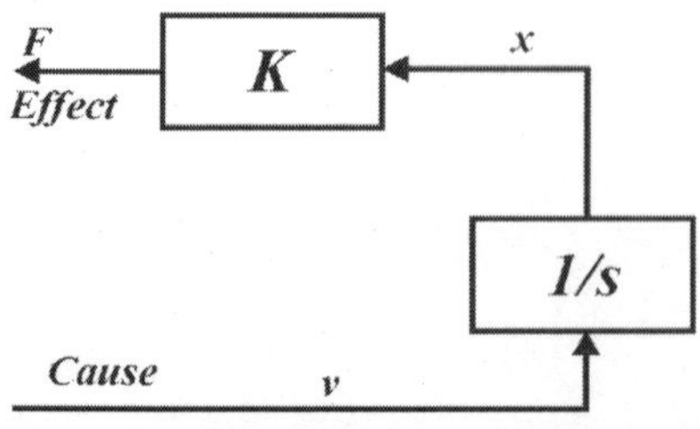

Figure 2.7: *Flow causes effort*

Table 2.1 shows a comparison between the different general systems indicating the effort and flow for each system

Table 2.1: *Comparison between general systems*

General system	**Effort** (e)	**Flow** (f)	$\int (f)dt$	$\int (e)dt$
Mechanical	Force (F)	Velocity (v)	Displacement (x)	Momentum (L)
Mechanical	Torque (T)	Angular velocity (ω)	Angle (θ)	Angular momentum (H)
Electrical	Voltage (V)	Current (i)	Charge (q)	Flux (ϕ)
Fluid	Pressure (P)	Flow rate (Q)	Volume (V)	Pressure momentum (Φ)

2.5 Modelling of a piston

2.5.1 First order transfer function

A fluid power piston without mass effect can be modelled as shown in Figure 2.8 where, k is the stiffness of fluid, C is the viscous damping between the moving piston and the fixed cylinder, x is the input displacement motion of the fluid inside the cylinder and y is the output displacement of the piston relative to the cylinder. The fixed cylinder here is considered as an inertial fixed reference (datum). The fluid is under a pressure P and acts at a piston area A. This system can be represented by the block diagram shown in Figure 2.9.

The resultant of the forces acting on the spring damper system shown in Figure 2.9 is equal to zero

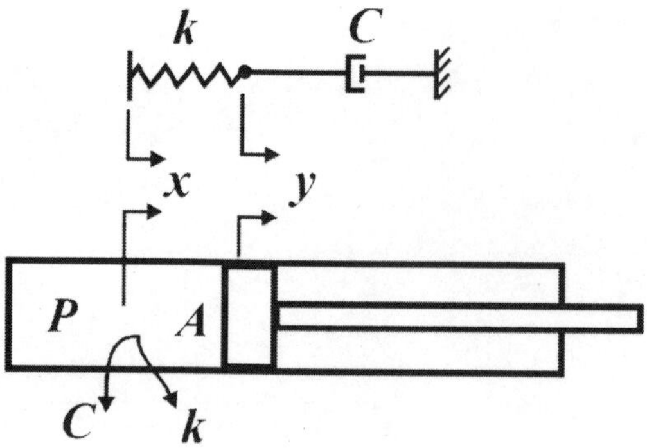

Figure 2.8: *First order dynamic model of a piston*

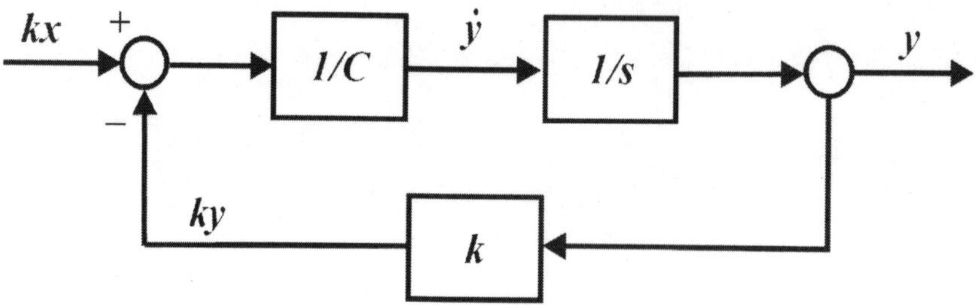

Figure 2.9: *First order block diagram of a piston*

$$kx - ky - C\dot{y} = 0 \tag{2.10}$$

or

$$kx - ky = C\frac{dy}{dt}$$

Assume having a unit input $x = 1$, then

$$C\frac{dy}{dt} = k(1 - y)$$

leading to the first order differential equation

$$dy = \frac{k}{C}(1 - y)dt$$

or

$$\frac{1}{1-y}dy = \frac{k}{C}dt$$

Integrating both sides of the equation

$$\int \frac{1}{1-y}dy = \int \frac{k}{C}dt$$

$$ln(1-y) = \frac{k}{C}t$$

Solving the equation for the output y

$$y = 1 - e^{-\frac{k}{C}t}$$

Thus, finding the output y in time domaine

$$y = 1 - e^{-t/\frac{C}{k}} = y = 1 - e^{-t/\tau}$$

where, $\tau = C/k$ is the time constant.
If the input is a sinusoidal harmonic motion at a frequency ω and an amplitude X, the output will be a harmonic motion at a new amplitude Y and a phase shift ϕ.

Input $x(t) = X sin(\omega t)$
Output $y(t) = Y sin(\omega t + \phi)$

Referring back to Figure 2.9, the summation of forces reads

$$x = \frac{C}{k}\dot{y} + y$$

In Laplace transform,

$$x = \frac{C}{k}sy + y$$

where s is the Laplace variable and can be related to the frequancy in the Real-Imaginary plane by $s = j\omega$. j here is the imaginary number $j = \sqrt{-1}$.

$$x = \frac{C}{k} j\omega y + y$$

or

$$x = y(\frac{C}{k} j\omega + 1)$$

The transfer function between x as an input and y as an output is then given by

$$\frac{y}{x} = \frac{1}{1 + (C/k)j\omega}$$

It has been shown that $C/k = \tau$ and $j\omega = s$. Back to Laplace transform

$$\frac{y(s)}{x(s)} = \frac{1}{1 + \tau s}$$

This transfer function represents a first order low-pass filter with a pole at $s = -1/\tau$. The Bode plot for this transfer function is shown in Figure 2.10. The magnitude is calculated in decibles (dB) where

$$dB = 20 \times log_{10}|\frac{y}{x}|$$

and the phase angle ϕ is in degrees where

$$\phi = tan^{-1}(-\tau\omega)$$

It is noticed that the roll-off of the transfer function at high frequency is equal to $-20dB/decade$ which corresponds to the first order frequency response function.

2.5.2 Second order transfer function

To build a more convinient model, a fluid power piston with mass effect can be modelled as shown in Figure 2.11, where m is the mass of the piston rod and suspended load, k is the stiffness of fluid, C is the viscous

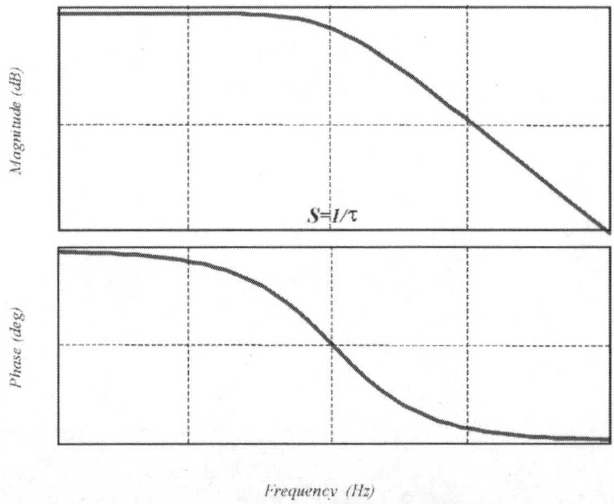

Figure 2.10: *Frequency response function of first order low-pass filter*

damping between the moving piston and the cylinder, x is the input displacement motion of the fluid inside the fixed cylinder and y is the output displacement of the piston relative to the fixed datum (cylinder). The fluid is under a pressure P and acts at a piston area A. This system can be represented by the block diagram shown in Figure 2.12.

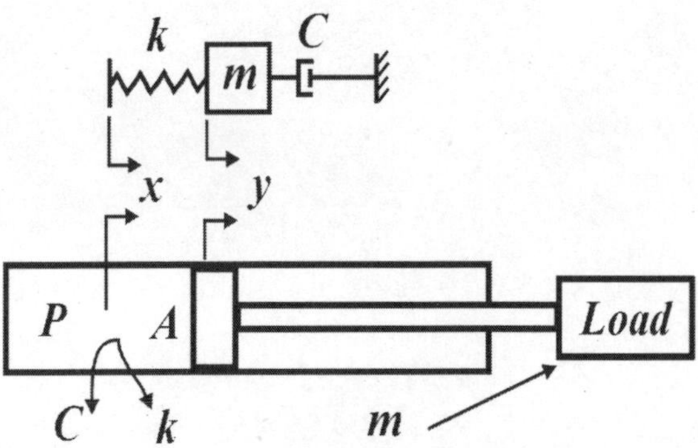

Figure 2.11: *Second order dynamic model of a piston*

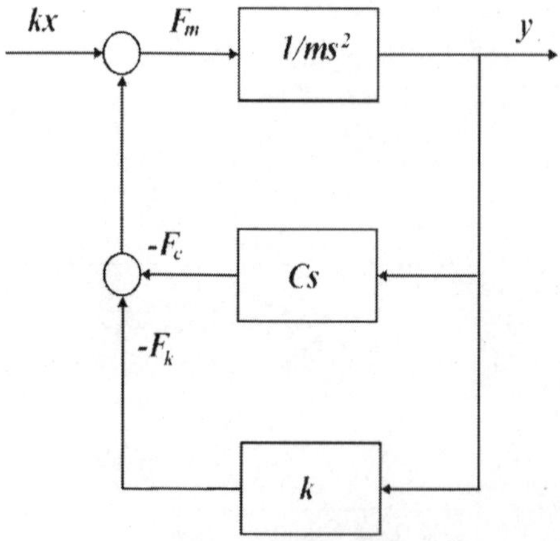

Figure 2.12: *Second order block diagram of a piston*

Looking at this system, one can determine the forces acting on the system:

Inertia force

$$F_m = m\ddot{y} = ms^2y$$

Damping force

$$F_C = C\dot{y} = -Csy$$

Spring force

$$F_k = k(x - y)$$

Applying Newton's law for the summation of forces results in

$$kx - F_C - F_k = F_m$$

Substituting the vlues of the forces givs

$$ms^2y + Csy + ky = kx \tag{2.11}$$

or

$$(ms^2 + Cs + k)y = kx$$

The transfer function between x and y is

$$\frac{y}{x} = \frac{k}{ms^2 + Cs + k} \tag{2.12}$$

or

$$\frac{y}{x} = \frac{k/m}{s^2 + (C/m)s + (k/m)}$$

but it known that the natural frequency $\omega_n = \sqrt{k/m}$ and the damping ratio $\xi = C/2m\omega_n$, then

$$\frac{y}{x} = \frac{\omega_n^2}{s^2 + 2\xi\omega_n s + \omega_n^2} \tag{2.13}$$

This transfer function represents a second order low-pass filter with a natural frequency ω_n and a damping ratio ξ. Figure 2.13 shows a Bode plot of the transfer function where the overshoot on the natural frequency is determined by the damping ratio ξ . The roll-off at high frequency is $-40dB/decade$ which gives a higher vibration isolation performance [5].

2.6 State space approach

The block diagram in Figure 2.14 depicts a state space modelling of a general dynamic system. A system can be represented by a set of first order linear differential equations [3]:

$$\dot{e} = Ax + Bu \tag{2.14}$$

$$y = Cx + Du \tag{2.15}$$

where

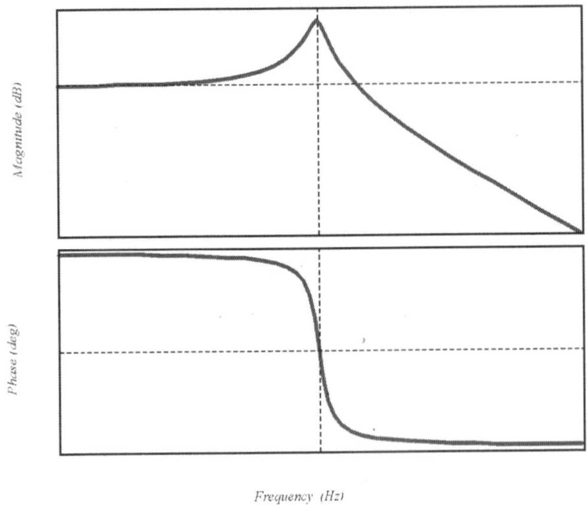

Figure 2.13: *Frequency response function of second order Low-pass filter*

$u =$ input vector.
$y =$ output vector.
$x =$ state vector.
$A =$ system matrix.
$B =$ input matrix.
$C =$ output matrix.
$D =$ feedthrough matrix.

The state vector is not unique and can be selected from the variables that can give information about the system. The simple oscillator shown in Figure 2.14 consists of a mass m, a spring k and a damping coefficient C.

The second order equation of motion of the simple oscillator reads

$$\ddot{x} + 2\xi\omega_n\dot{x} + \omega_n^2 x = \frac{1}{m}f \tag{2.16}$$

The state variables can be selected as the displacement x and the velocity $\dot{x}$ as follows

$$e_1 = x$$

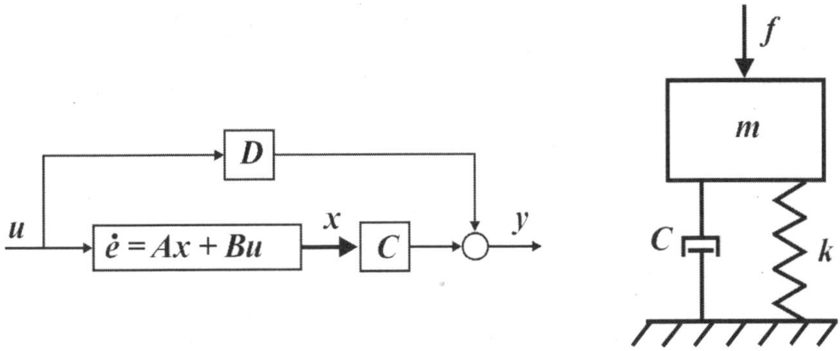

Figure 2.14: *State space representation of a simple oscillator*

$$e_2 = \dot{x}$$

then the derivatives of the state variables are

$$\dot{e}_1 = \dot{x} = e_2$$

$$\dot{e}_2 = \ddot{x} = -2\xi\omega_n\dot{x} - \omega_n^2 x + \frac{1}{m}f$$

In matrix form

$$\begin{pmatrix} \dot{e}_1 \\ \dot{e}_2 \end{pmatrix} = \begin{pmatrix} \dot{x} \\ \ddot{x} \end{pmatrix} = \overbrace{\begin{pmatrix} 0 & 1 \\ -\omega_n^2 & 2\xi\omega_n \end{pmatrix}}^{A} \begin{pmatrix} x \\ \dot{x} \end{pmatrix} + \overbrace{\begin{pmatrix} 0 \\ \frac{1}{m} \end{pmatrix}}^{B} f \tag{2.17}$$

The system matrix here reads

$$A = \begin{pmatrix} 0 & 1 \\ -\omega_n^2 & 2\xi\omega_n \end{pmatrix}$$

and the input matrix reads

$$B = \begin{pmatrix} 0 \\ \frac{1}{m} \end{pmatrix}$$

If the measured output is the displacement x, $y = e_1 = x$, then the output matrix reads

$$C = \begin{pmatrix} 1 & 0 \end{pmatrix}$$

If the measured output is the velocity $\dot{x}$, $y = e_2 = \dot{x}$, then the output matrix reads

$$C = \begin{pmatrix} 0 & 1 \end{pmatrix}$$

In these two cases, the feedthrough matrix D is a zero matrix $D = (0)$ which means that there is no direct influence of the input on the output. If the measured output is the acceleration, $y = \dot{e}_2 = \ddot{x}$, then the feethrough matrix $D = 1/m$ and the output equation reads

$$y = \underbrace{\begin{pmatrix} -\omega_n^2 & -2\xi\omega_n \end{pmatrix}}_{C} \begin{pmatrix} x \\ \dot{x} \end{pmatrix} + \underbrace{\frac{1}{m}}_{D} f$$

This shows that the output matrix is

$$C = \begin{pmatrix} -\omega_n^2 & -2\xi\omega_n \end{pmatrix}$$

and the feedthrough matrix is

$$D = \frac{1}{m}$$

Example 2.2

Consider the two piston system shown in Figure 2.15. The inputs to the system are the two hydraulic forces F_1 and F_2 and the outputs are the two displacements x_1 and x_2. m_1 and m_2 are the two masses of the loads with the connected rods. k_1 and C_1 are the hydraulic stiffness and the viscous damping of piston (1) respectively and k_2 and C_2 are the hydraulic stiffness and the viscous damping of piston (2) respectively. Determine the state space matrices A, B, C and D in symbolic form. Suggest adequate numbers for the masses, the stiffnesses and the damping coefficients and draw the Bode plots for the transfer functions x_1/F_1 and x_2/F_2.

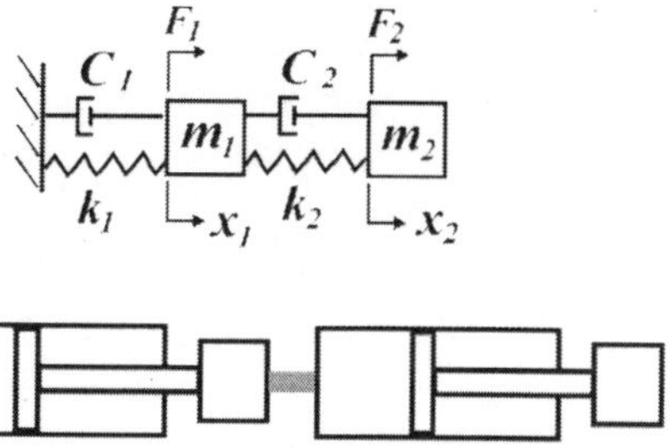

Figure 2.15: *Two piston system for example 2.2*

Solution

This is a two degrees of freedom system that can be represented by two second order differential equations

$$m_1\ddot{x}_1 = F_1 - k_1x_1 - C_1\dot{x}_1 + k_2(x_2 - x_1) + C_2(\dot{x}_2 - \dot{x}_1)$$

$$m_2\ddot{x}_2 = F_2 - k_2(x_2 - x_1) - C_2(\dot{x}_2 - \dot{x}_1)$$

Solving for the accelerations $\ddot{x}_1$ and $\ddot{x}_2$

$$\ddot{x}_1 = \frac{1}{m_1}F_1 + x_1\left(\frac{-k_1 - k_2}{m_1}\right) + \dot{x}_1\left(\frac{-C_1 - C_2}{m_1}\right) + x_2\left(\frac{k_2}{m_1}\right) + \dot{x}_2\left(\frac{C_2}{m_1}\right)$$

$$\ddot{x}_2 = \frac{1}{m_2}F_2 + x_1\left(\frac{k_2}{m_2}\right) + \dot{x}_1\left(\frac{C_2}{m_2}\right) + x_2\left(\frac{-k_2}{m_2}\right) + \dot{x}_2\left(\frac{-C_2}{m_2}\right)$$

Now building the state space equations from the equations of motion

$$\begin{pmatrix} \dot{x}_1 \\ \ddot{x}_1 \\ \dot{x}_2 \\ \ddot{x}_2 \end{pmatrix} = \begin{pmatrix} 0 & 1 & 0 & 0 \\ \frac{-k_1-k_2}{m_1} & \frac{-C_1-C_2}{m_1} & \frac{k_2}{m_1} & \frac{C_2}{m_1} \\ 0 & 0 & 0 & 1 \\ \frac{k_2}{m_2} & \frac{C_2}{m_2} & \frac{-k_2}{m_1} & \frac{-C_2}{m_1} \end{pmatrix} \begin{pmatrix} x_1 \\ \dot{x}_1 \\ x_2 \\ \dot{x}_2 \end{pmatrix} + \begin{pmatrix} 0 & 0 \\ \frac{1}{m_1} & 0 \\ 0 & 0 \\ 0 & \frac{1}{m_2} \end{pmatrix} \begin{pmatrix} F_1 \\ F_2 \end{pmatrix}$$

The outputs are the displacements x_1 and x_2

$$\begin{pmatrix} x_1 \\ x_2 \end{pmatrix} = \begin{pmatrix} 1 & 0 & 0 & 0 \\ 0 & 0 & 1 & 0 \end{pmatrix} \begin{pmatrix} x_1 \\ \dot{x}_1 \\ x_2 \\ \dot{x}_2 \end{pmatrix} + \begin{pmatrix} 0 & 0 \\ 0 & 0 \end{pmatrix} \begin{pmatrix} F_1 \\ F_2 \end{pmatrix}$$

These four state space matrices can used to find the Bode plots. The two bode plots representing x_1/F_1 and x_2/F_2 are shown in Figure 2.16 and Figure 2.17, respectively

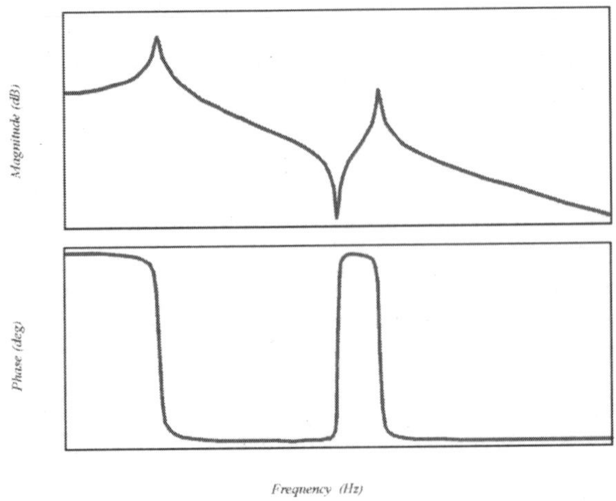

Figure 2.16: *Bode plot for the transfer function* x_1/F_1

2.7 Modelling of rotating elements

For angular motion, rotary actuators are needed, there are several types of rotary actuators; gear type, piston type and vane type. When the actuator is a rotary element, the equation of motion reads

$$I\ddot{\theta} + C_\theta\dot{\theta} + k_\theta\theta = T \tag{2.18}$$

or

$$I\alpha + C_\theta\omega + k_\theta\theta = T \tag{2.19}$$

where

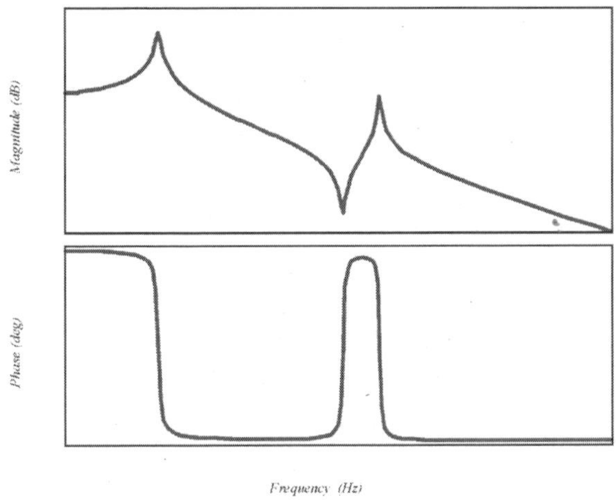

Figure 2.17: *Bode plot for the transfer function* x_2/F_2

I = Moment of inertia.
T = Torque
k_θ = Angular stiffness
C_θ = Angular damping
θ = Angle of rotation
$\omega = \dot{\theta}$ = Angular velocity
$\alpha = \ddot{\theta}$ = Angular acceleration

This model can be treated as the previous one, either by finding transfer functions or state space approach. The torque T can be calculated from

$$T = PAR$$

where A is the area on which the pressure P acts and R is the distance from the center of pressure to the center of rotation.

2.8 Modelling of control valves

A control valve is the main element in fluid power system. The most common control valve is the spool valve shown in Figure 2.18. It consists of a spool with circular cross section slides inside a cylindrical casing. The spool is made of lands connected together by central rods. The general equation of motion of this type of valves can be represented by the following equation of motion, [6], [7]

$$P_1 A_1 - P_2 A_2 = m\ddot{x} + C\dot{x} + kx \tag{2.20}$$

where

P_1 = Pilot pressure acting on the left land of the spool.
P_2 = Pilot pressure acting on the right land of the spool.
A_1 = Cross sectional area of the left land of the spool.
A_2 = Cross sectional area of the right land of the spool.
m = The mass of the sliding spool.
C = The viscous damping coefficient in the valve.
k = The stiffness factor of the fluid.

The pressure values included in equation (2.20) are the low pilot pressure values of the pilot fluid used to move the spool and not the high system pressure used to move the load or the piston. Looking at the port from which fluid passes in or out of the valve with high pressure, the relationship between the flow rate and the pressure which represents the steady state response can be determined as follows

$$Q = C_d A \sqrt{\frac{2\Delta P}{\rho}} \tag{2.21}$$

where

Q = The fluid flow rate through the port.
ΔP = The difference of high pressure passing through the valve.
C_d = The discharge coefficient of the orifice (port).
A = Cross sectional area of the port.
ρ = The fluid density.

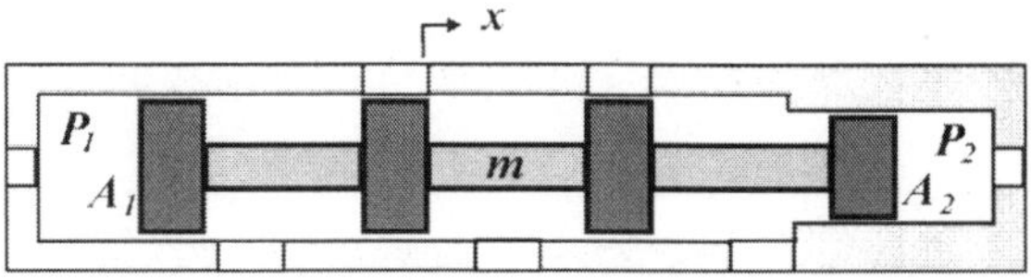

Figure 2.18: *General design of spool valve*

2.9 Further discussion

So far, all the discussed different models have been considered as linear models. The rapid change of a nonlinear load accompanied with the need for a variable displacement pump that leads to variable pressure can be a strong reason for the nonlinearity of fluid power systems. Furthermore, nonlinearity is added to the system due to dry friction between the sliding pistons and the casing cylinders. This nonlieartity led some of the researchers to use nonlinear control techniques like H-infinity, neural networks and Fyzzy logic control to increase the control performance and robustness.

For more information about nonlinear systems and nonlinear control techniques, the reader can refer to some advance researches like those in [8], [9].

2.10 References

[1] F. Yeaple, *Fluid Power Design Handbook*, Third Ed., Marcel Dekker,1996.

[2] K. Ogata, *Modern Control Engineering*, Prentice Hall, 2002.

[3] G. Franklin, *Feedback Control of Dynamic Systems*, Third Ed., Addison-Wesley, 1994.

[4] M. Pinches and J. Ashby, *Power Hydraulics*, Prentice Hall, 1989.

[5] A. Preumont, *Vibration Control of Active of Active Structures, an Introduction*, Second Ed., Kluwer Academic Publishers, 2002.

[6] X. Wang, Q. Yang and G. Zheng, Dynamic Modeling and Active

Decoupling Control of Octo-pneumatic Actuator Vibration Isolation Platform, *Chinese Journal of Mechanical Engineering*, Volume 22, Number 1, 2009.

[7] T. Lim, D. Zhang, S. Sheng, L. Eicher and W. Lau, Control of Hybrid Magnetic Bearing for Blood Pumps with PID and Robust Model Reference Adaptive Control Scheme, *Mechanical Systems and Signal Processing*, Volume 23, Issue 7, 2009.

[8] W. Kemmetmuller, Nonlinear Pressure Control of Self-supplied Variable Displacement Axial Piston Pumps, *IEEE Transactions on Mechatronics*, Volume 12, Issue 1, 2007.

[9] C. Chen, L. Lieu, P. Liao and G. Chiu, Fuzzy Controller Design for Synchronous Motion in a Dual-cylinder Electro-hydraulic Systems, *Control Engineering Practice*,, Volume 16, Issue 6, 2008.

Chapter 3

Control Systems

Fluid power systems stay useless unless they are organized and controlled to be directed to specific mechanical and industrial applications. Control techniques depend strongly on the deep understanding of the system and the type of fluid used [1]. Control system can be one the following:

- **mechanical:** by utilizing the fluid itself at low pressure to actuate the control valves that in turn control the high pressure fluid and pass it to move the actuators at the application.
- **Electrical:** that can be analog or digital. This way of control utilizes the signals measured by sensors near the load or any signal fed directly to actuate linear or rotary electric motors whose motion is used to actuate the control valve. Here, the control valve is also used to move the end effecting actuators.

3.1 Servo Control systems

Usually, the control systems include both electrical and mechanical components. The most common control system is the *Servo* control system. A control system is said to be servo if the input signal is amplified (*Conditioning*) and if there is a signal fed back from the output to the input in a closed-loop *Feedback*. The closed-loop feedback system shown in Figure 3.1 is usually called $follow - up$ system that aims at keeping the output at a given value by minimizing the error.

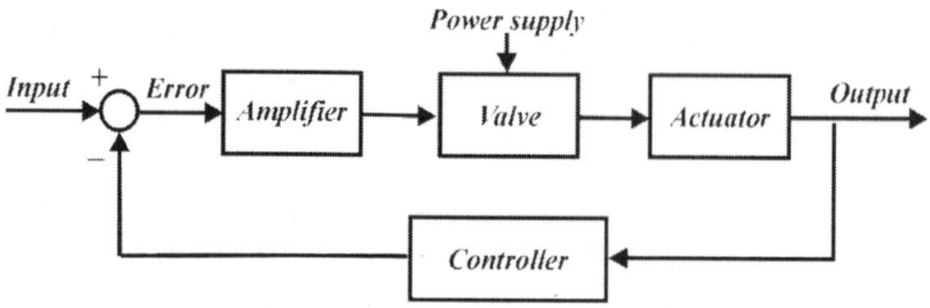

Figure 3.1: *Block diagram of closed-loop feedback servo system*

A follow-up hydraulic closed-loop feedback servo system undergos two types of systems:

1. Valve operated servo control system: used for lower power applications.

2. Pump operated servo control system: used for large power applications.

3.1.1 Valve operated servo control

A valve operated servo control system is usually used for lower power applications. Valve operated system is inherently inefficient. It operates with a constant or variable displacement pump and the valve acts as an orifice which increases the resistance in the system increasing the temperature and decreasing the performance [2]. Generally, a valve operated servo control system has the following characteristics:

- The circuit is easily designed and constructed with simple components.

- It has a rapid dynamic response because of having lower inertia.

- One single pump can be enough to give power for the whole system whereas valves are distributed amongst the different actuators and applications.

One of the most important concepts in valves is the valve lap where the spool valve can have zero lap, underlap or overlap as shown in Figure 3.2.

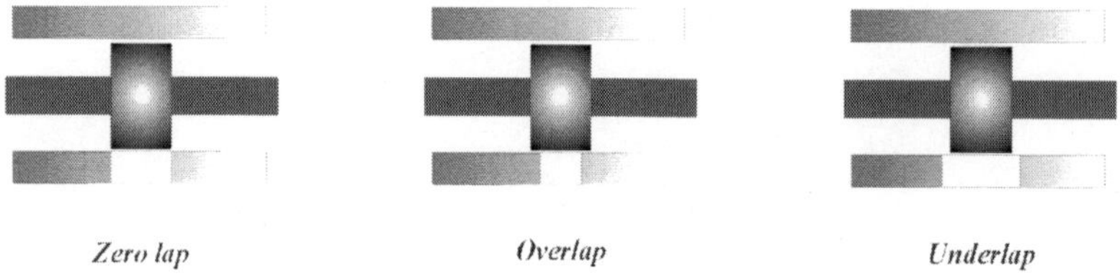

Figure 3.2: *Valve lap*

- *Zero lap:* the width of the spool land is exactly the same as the width of the port which leads to a proportional relation between the displacement of the spool and the quantity of fluid flowing through the valve. This is practically difficult to manufacture and not easy to obtain.

- *Overlap:* the width of the land is bigger than that of the port which leads to having a dead zone where there is a movement of the spool without having any flow in this region.

- *Underlap:* the width of the land is smaller than that of the port leading to a continuation in flow even when the spool land is on the mid point of the port (null point).

Since it is not easy to manufacture a zero lap spool because of the need to high accuracy, this design is usually replaced by an overlap with notches on the edge of the land. The underlap design is usefull in achieving a high response at null point and is used to compensate for the loss in fluid when there is a leakage in the system.
The discharge equation in the valve reads

$$Q = C_d A \sqrt{\frac{2\Delta P}{\rho}} \tag{3.1}$$

where

$Q =$ The fluid flow rate through the port.
$\Delta P =$ The difference of high pressure passing through the valve.
$C_d =$ The discharge coefficient of the orifice (port).

$A =$ Cross sectional area of the port.
$\rho =$ The fluid density.

Example 3.1

Consider the valve srvo copy machine shown in Figure 3.3. Determine the transfer function between the stylus displacement u and the tool holder displacement y.

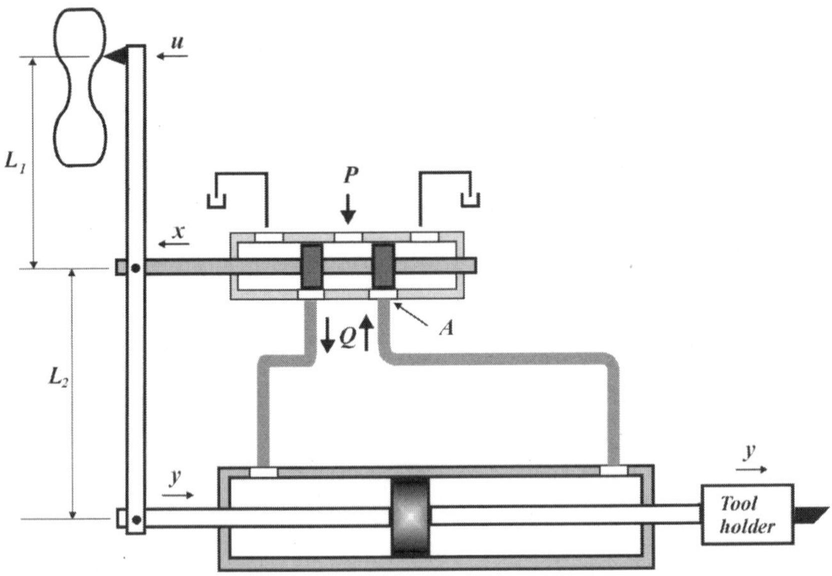

Figure 3.3: *Valve servo Copy mechanism for example 3.1*

Solution

The system sonsists of a spool valve and a douple acting piston. Both, the spool of the valve and the rod of the piston are connected to the feedback lever arm. The end of the lever arm is influenced by the template input displacement u leading to the tool holder machining displacement y passing through the intermediate valve spool displacement x.

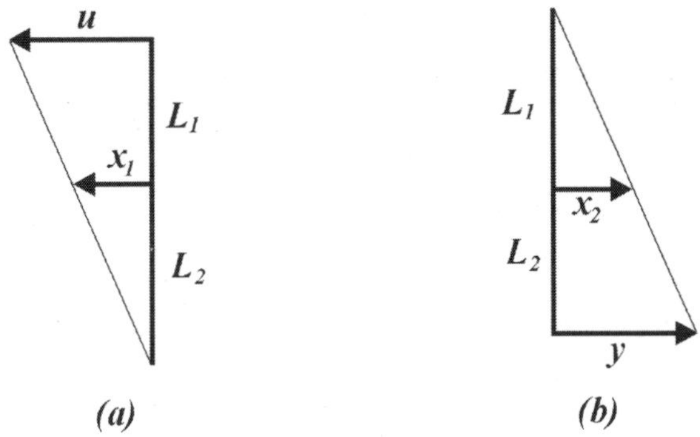

Figure 3.4: *Displacemenet triangles for example 3.1*

Looking at Figure 3.4(a), from similar triangles

$$\frac{x_1}{L_2} = \frac{u}{L_1 + L_2}$$

solving for x_1

$$x_1 = \frac{uL_2}{L_1 + L_2}$$

Looking at Figure 3.4(b), from similar triangles

$$\frac{x_2}{L_1} = \frac{y}{L_1 + L_2}$$

solving for x_2

$$x_2 = \frac{yL_1}{L_1 + L_2}$$

The spool displacement x

$$x = x_1 - x_2 = \frac{uL_2}{L_1 + L_2} - \frac{yL_1}{L_1 + L_2}$$

for a special case when $L_1 = L_2$

$$x = \frac{u - y}{2}$$

If there is no leakage in the flow considering a zero lap servo valve, the discharge equation reads

$$Q = C_d x d \sqrt{\frac{2\Delta P}{\rho}}$$

where d is the average width of the port (area of the port is $A = xd$). For incompressible hydraulic fluid with a constant pressure input to the valve

$$Q = K_v x$$

where $K_v = C_d d\sqrt{2\Delta P/\rho}$ is the valve constant at specific constant pressure and specific density. On the other hand, the flow rate from the valve to the piston is calculated from

$$Q = A\frac{dy}{dt}$$

Equating the two equations of flow,

$$K_v x = A\frac{dy}{dt}$$

or

$$K_v \frac{u - y}{2} = A\frac{dy}{dt}$$

Applying Laplace transform with zero initial conditions

$$K_v(u(s) - y(s)) = 2Asy(s)$$

using separation of variables to solve the first order equation

$$K_v u(s) = y(s)(K_v + 2As)$$

Solving for the transfer function y/u

$$\frac{y(s)}{u(s)} = \frac{K_v}{(K_v + 2As)}$$

or

$$\frac{y(s)}{u(s)} = \frac{1}{(1 + \frac{2A}{K_v}s)}$$

$$\frac{y(s)}{u(s)} = \frac{1}{(1 + \tau s)}$$

This transfer function is a first order frequency response function with a time constant $\tau = 2A/K_v$. The Bode plot for this transfer function is shown in Figure 3.5.

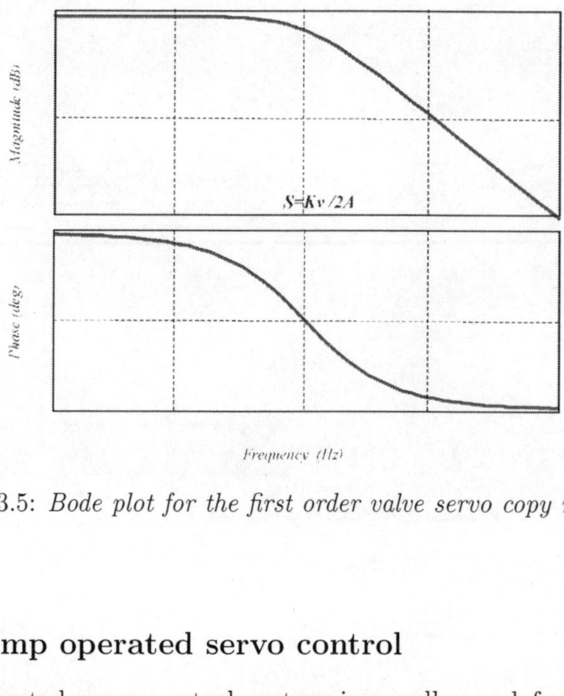

Figure 3.5: *Bode plot for the first order valve servo copy machine*

3.1.2 Pump operated servo control

A pump operated servo control system is usually used for large power applications especially hydraulic rotating motors. It operates with a variable displacement pump where the system is controlled by changing the displacement of the pump [3]. A pump operated servo control system has the following general characteristics:

- The circuit is compact with more complicated components.
- It gives much higher power to drive high inertia loads.
- Friction losses are minimized in this system which increases the efficiency of the system.
- Used in general to drive hydraulic motors (Hydrostatic transmission system).

The pump operated servo control system is usually used for hydrostatic power transmission systems where a variable displacement pump with a displacement y driven by an electric motor or internal combustion engine at a speed Ω_p, delivers hydraulic oil with a flow rate of q to rotate a hydraulic motor at a speed of Ω_m.

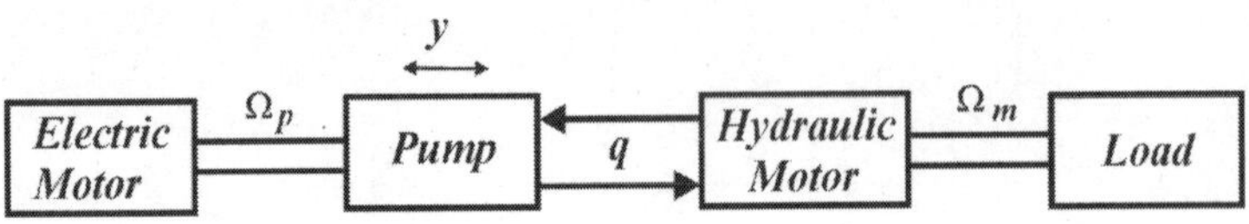

Figure 3.6: *Block diagram of a hydrostatic transmission system*

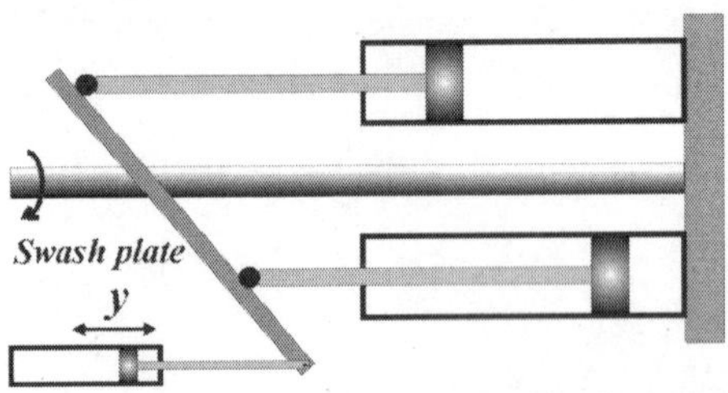

Figure 3.7: *Section of a reciprocating variable displacement pump*

Consider the hydrostatic transmission system shown in Figure 3.6 with a reciprocating positive variable displacement pump shown in Figure 3.7. Assuming there is neither leakage nor compressibility in the system

Flow from the pump = Flow to the motor

$$q_p = q_m = q$$

$$\Omega_p d_p = \Omega_m d_m$$

where $\Omega_p = constant$ is the constant angular speed of the pumps rotor and d_m is the displacement of the motor per radian. If the control piston in the pump moves a displacement y as shown in Figure 3.7, the flow rate of the pump is proportional to that displacement

$$q_p = K_p y$$

where K_p is the pump flow constant at a constant speed, this leads to

$$q_p = q_m = K_p y = \Omega_m d_m$$

Hence, finding the transfer function between the pump's control displacement and the speed of the motor

$$\frac{\Omega_m}{y} = \frac{K_p}{d_m} \tag{3.2}$$

This gives a constant relationship between the input and the output without being influenced by the dynamics of the system which is too much ideal because the maximum speed cannot be achieved instantaneously at zero time.

Leakage effect

In real hydraulic systems, leakage occurs in pumps and hydraulic motors which includes a significant influence on the flow rate in the system. The leakage coefficients of the pump and motor are denoted by λ_p and λ_m, respecively. The leakage coeeficients are proportional to the pressure value in the system [4].

Pump leakage = $\lambda_p P_p$.
Actual flow in pump = $K_p y - \lambda_p P_p = q$.

Motor leakage $= \lambda_m P_m$.
Actual flow to motor $= q - \lambda_m P_m = \Omega_m d_m$

so

$$\Omega_m d_m = K_p y - \lambda_p P_p - \lambda_m P_m$$

If the combined leakage coefficient λ for both pump and motor is

$$\lambda = \lambda_p + \lambda_m$$

and the pressure drop is negligible, such that

$$P = P_p = P_m$$

then

$$\Omega_m d_m = K_p y - \lambda P \tag{3.3}$$

Equating the output mechanical power to the input hydraulic power in the motor reads

$$Power = T_m \Omega_m = P_m q_m$$

where T_m is the torque output of the motor

$$T_m \Omega_m = P_m \Omega_m d_m$$

therefore,

$$T_m = P_m d_m$$

but the torque can be related to the inertia I and the angular acceleration α by

$$T_m = \alpha I = \frac{d\Omega_m}{dt} I$$

thus

$$P_m d_m = \frac{d\Omega_m}{dt} I$$

Manipulating Equation (3.3) gives

$$\Omega_m d_m = K_p y - \lambda P$$

$$P = \frac{K_p y - \Omega_m d_m}{\lambda}$$

multiplying both sides by d_m gives

$$P d_m = \frac{d_m K_p y - \Omega_m d_m^2}{\lambda}$$

therefore,

$$T_m = P d_m$$

or

$$I\frac{d\Omega_m}{dt} = \frac{d_m}{\lambda}(K_p y - \Omega_m d_m)$$

Using Laplace transform with zero initial conditions

$$Is\Omega_m(s) = \frac{d_m}{\lambda}(K_p y(s) - \Omega_m(s) d_m)$$

$$\Omega_m(s)\left(Is + \frac{d_m^2}{\lambda}\right) = \frac{d_m}{\lambda} K_p y(s)$$

Solving for the transfer function between $y(s)$ and $\Omega_m(s)$

$$\frac{\Omega_m(s)}{y(s)} = \frac{d_m K_p/\lambda}{Is + d_m^2/\lambda}$$

multiplying numerator and denominator by λ/d_m^2 gives

$$\frac{\Omega_m(s)}{y(s)} = \frac{K_p}{d_m}\left(\frac{1}{1 + (\lambda I/d_m^2)s}\right) \tag{3.4}$$

This represents a first order low-pass filter

$$\frac{\Omega_m(s)}{y(s)} = \frac{K_p}{d_m}\left(\frac{1}{1 + \tau s}\right)$$

with a time constant $\tau = \lambda I/d_m^2$. The bode plot of this transfer function is shown in Figure 3.8.

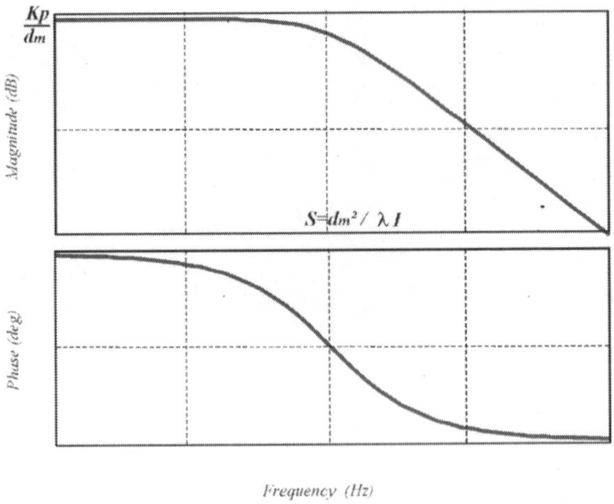

Figure 3.8: *Frequency response function $\Omega_m(s)/y(s)$ with leakage effect*

Compressibility effect

Hydraulic oils are considered, ideally, incompressible for simplicity of calculations. However, there is a small amount of compressibility in these fluids that introduces some flexibility causing dynamic influence on the system.

In solid mechanics, the modulus of elasticity is considered as the measure of flexibility of solid materials. Similarly, flexibility of fluids is measured by a constant called *Bulk modulus B* and can be defined by

$$B = \frac{Volumetric\ stress}{Volumetric\ strain}$$

or

$$B = \frac{P}{\Delta V/V}$$

where P is the presure and V is the original volume of fluid between the pump and the motor. Compressibility has a direct influence on the pressure. Although, it causes a loss in the flow q_c, where

$$q_c = \frac{d}{dt}\Delta V = \frac{d}{dt}\left(\frac{VP}{B}\right) = \left(\frac{V}{B}\right)\frac{d}{dt}P$$

The actual flow reches to the motor is

$$\Omega_m d_m = \underbrace{K_p y}_{Pump\ delivery} - \underbrace{\lambda P}_{Leakage\ loss} - \underbrace{\left(\frac{V}{B}\right)\frac{d}{dt}P}_{Compressibility\ loss} \tag{3.5}$$

From previous discussion, it is known that

$$Pd_m = T_m = I\frac{d}{dt}\Omega_m$$

or

$$P = \frac{I}{d_m}\frac{d\Omega_m}{dt}$$

Substituting back in eauqtion (3.5) results in

$$\Omega_m d_m = K_p y - \lambda\left(\frac{I}{d_m}\frac{d\Omega_m}{dt}\right) - \frac{V}{B}\frac{d}{dt}\left(\frac{I}{d_m}\frac{d\Omega_m}{dt}\right)$$

or

$$\Omega_m d_m = K_p y - \lambda\left(\frac{I}{d_m}\frac{d\Omega_m}{dt}\right) - \frac{VI}{Bd_m}\frac{d^2\Omega_m}{dt^2}$$

Using Laplace transfrom

$$\Omega_m(s)d_m = (K_p y(s)) - \left(\frac{\lambda I}{d_m}s\Omega_m(s)\right) - \left(\frac{VI}{Bd_m}s^2\Omega_m(s)\right)$$

Manipulating

$$\Omega_m(s)\left(d_m + \left(\frac{\lambda I}{d_m}\right)s + \left(\frac{VI}{Bd_m}\right)s^2\right) = K_p y(s)$$

Solving for the transfer function Ω_m/y gives

$$\frac{\Omega_m(s)}{y(s)} = \frac{K_p}{d_m + (\lambda I/d_m)\,s + (VI/Bd_m)\,s^2}$$

or

$$\frac{\Omega_m(s)}{y(s)} = \frac{K_p}{d_m}\left(\frac{Bd_m^2/VI}{s^2 + (\lambda B/V)s + (Bd_m^2/VI)}\right) \tag{3.6}$$

Which represents a second order low-pass filter. But the general characteristic transfer function reads

$$\frac{\Omega_m(s)}{y(s)} = \frac{K_p}{d_m}\left(\frac{\omega^2}{s^2 + 2\xi\omega s + \omega^2}\right) \tag{3.7}$$

Matching equations (3.6) and (3.7) gives the following parameters of the system

$$\omega^2 = \frac{Bd_m^2}{VI}$$

and

$$2\xi\omega = \frac{\lambda B}{V}$$

where ω is the undamped natural frequency and ξ is the damping ratio in the system. From the previous comparison, the natural frequency reads

$$\omega = \sqrt{\frac{Bd_m^2}{VI}}$$

It is important for any hydraulic system to increase the stiffness and push the natural frequency to the highest possible value. The following steps can be taken into account to increase the natural frequency:

- Increasing the motor displacement d_m by increasing the pump flow rate.
- Decreasing the original volume V of the fluid between the pump and the motor. This is possible by installing the motor close enough to the pump.
- Minimizing the load inertia.

Knowing the natural frequency of the system, the hydraulic stiffness can be determined as follows: for a revolution of 1 radian, the volume difference is equal to the motor displacement $\Delta V = d_m$, therefore

$$P_m = \frac{Bd_m}{V}$$

and the torque of the motor is

$$T_m = d_m P_m = d_m \left(\frac{Bd_m}{V} \right) = \frac{Bd_m^2}{V}$$

and can be calculated from

$$T_m = K_H \times \theta$$

where K_H is the hydraulic stiffness and θ is the angle of rotation. But $\theta = 1$ radian, then $K_H = T_m$

$$K_H = \frac{Bd_m^2}{V} \tag{3.8}$$

Solving for the natural frequency

$$\omega = \sqrt{\frac{K_H}{I}} = \sqrt{\frac{Bd_m^2}{VI}}$$

The damping ratio can be calculated from the previous parameters of the system as follows

$$2\xi\omega = \frac{\lambda B}{V}$$

substituting the value of the natural frequency gives

$$\xi = \sqrt{\frac{\lambda^2 BI}{4Vd_m^2}} \tag{3.9}$$

It is clear from the previous discussion that the damping ratio depends strongly on the leakage in the system. Knowing the different parameters of the system, the transfer function in equation (3.6) can be presented using the Bode plot shown in Figure 3.9.

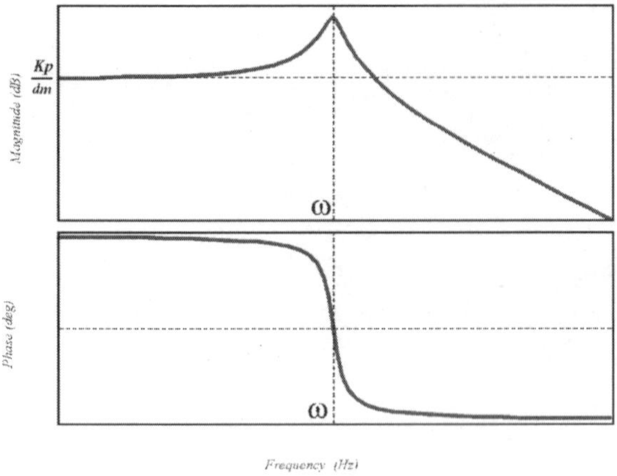

Figure 3.9: *Frequency response function $\Omega_m(s)/y(s)$ with leakage and compressibilty effects*

Example 3.2

Consider a reversible hydrostatic transmission system with a variable displacement pump and a constant displacement hydraulic motor with the following characteristics:

Leakage coefficient of pump and motor $\lambda = 0.01$ l/min/bar.
Load inertia $I = 300\ N.m.s^2$.
The motor displacement $d_m = 25$ ml/radian.
The maximum motor speed $\Omega_m = 200$ rpm.
Motor acceleration $\alpha = 1.05\ rad/s^2$.
Pump speed $\Omega_p = 1400$ rpm.
Overall effeiciecy $\eta = 85\%$.
Pump control stroke $y = 0.1$ m.

Neglecting the compressibility effect, calculate:

1. System pressure (pipe friction losses are negligible).
2. The actual pump capacity.

3. The power of the electric motor needed to drive the pump.

4. The time constant τ and the frequency response function Ω_m/y.

Solution

1. The torque at the load

$$T_m = I\alpha$$
$$T_m = 300(Nms^2) \times 1.05 \ (rad/s^2)$$
$$T_m = 315 \ N.m$$

But the hydraulic motor torque is

$$T_m = P_m d_m$$
$$P_m = T_m/d_m = 315(Nm)/25 \times 10^{-6} \ (m^3)$$
$$P_m = 125.7 \times 10^5 Pa = 125.7 \ bar$$

2. The fluid flow rate reaching to the motor is

$$q_m = \Omega_m d_m \times 2\pi$$
$$q_m = 200(rev/min) \times 25 \times 10^{-6}(m^3) \times 2\pi$$
$$q_m = 31.4 \times 10^{-3} m^3/min = 31.4 \ l/min$$

The total leakage from the pump and the motor is

$$q_l = \lambda P_m$$
$$q_l = 0.01(l/min/bar) \times 125.7(bar)$$
$$q_l = 1.26 \ l/min$$

The total actual pump capacity is

$$q_p = q_m + q_l$$
$$q_p = 31.4 + 1.26 = 32.66\ l/min$$

3. Electric motor power in (kW) is

$$Electric\ motor\ power = \frac{flow(l/min) \times Prsssure\ (bar)}{600 \times Overall\ effeciency}$$

$$Electric\ motor\ power = \frac{32.66(l/min) \times 125.7(bar)}{600 \times 0.85} = 8.04\ kW$$

4. The time constant

$$\tau = \frac{\lambda I}{d_m^2}$$

$$\lambda = 0.01 \frac{l}{min.bar} \times \frac{1min}{60s} \times \frac{1bar}{10^5(N/m^2)} = \frac{1 \times 10^{-10}}{60}(m^5/N/s)$$

$$\tau = \frac{(10^{-10}/60)(300)}{(25 \times 10^{-6})^2} = 0.8\ s$$

The transfer function

$$\frac{\Omega_m}{y} = \frac{K_p}{d_m}\left(\frac{1}{1+\tau s}\right)$$

The pump coefficient

$$K_p = \frac{q_p}{y} = \frac{32.66}{60 \times 10^3} \times \frac{1}{0.1} = 5.44 \times 10^{-3}\ m^2/s$$

Therefore

$$\frac{\Omega_m}{y} = \frac{5.44 \times 10^{-3}}{25 \times 10^{-6}}\left(\frac{1}{1+0.8s}\right)$$

$$\frac{\Omega_m}{y} = 217.6\left(\frac{1}{1+0.8s}\right)$$

3.2 Fluid power symbols

It is not easy to represent fluid power circuits by technical drawing everytime it is required to explain a circuit. Therefore, it is important to use a simplified way of representation. Special symbols are used to represent every part of the circuit. These symbols are considered as a language code and can be understood by all the system engineers working in this field. To simplify utlizing these symbols, they are separated according to their usage in the circuit.

3.2.1 Symbols of lines

Table 3.1 shows the symbols used to draw lines and basic variation symbols. In this table, the solid straight line is used to represent the main flow solid pipe line while the solid curved line represents a flexible hose main flow line. The dotted line is used for the pilot low pressure line that brings the control signal and the dash-dotted line is used to enclose specific parts of the circuit for the puprpose of grouping. Note here that using inclined arrow means having variable parameters and using perpendicular arrow means having pressure compensation. The solid arrow head is usually used for hydraulic circuits while empty arrow head is used for pneumatic circuits.

3.2.2 Symbols of pumps, compressors and prime movers

Table 3.2 depicts a definition for the different power sources in fluid power systems. Single and reversible direction hydraulic pumps with fixed and variable flow control is shown besides to the symbols of the different prime movers (electric motors and internal combustion engines).

3.2.3 Symbols of Actuators

Some symbols used for the actuators are shown in Table 3.3. Symbols of rotary actuators or hydraulic motors are shown for single and reversible direction with fixed and variable displacement control. In linear actuators, it is possible to have a double lined symbol or a single lined symbols (simplified form). It is possible also to show if the piston is with single

Table 3.1: *Fluid power symbols (Lines)*

Symbol	Definition
	Main line
	Pilot line
	Enclosure outline
	Hydraulic flow direction
	Pneumatic flow direction
	Flexible pipe line
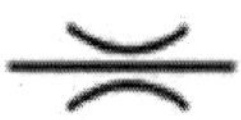	Constant flow restriction
	Variable flow restriction
	Pressure Compenation (small perpendicular arrow)
	Temperature effect
	Vented reservoir (hydraulic)

Table 3.2: *Fluid power symbols (Pumps, compressors and prime movers)*

Symbol	**Definition**
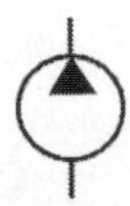	Single direction, fixed displacement pump
	Reversible, fixed displacement pump
	Reversible, variable displacement pump
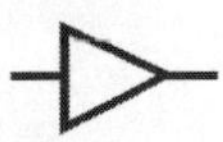	Air compressor
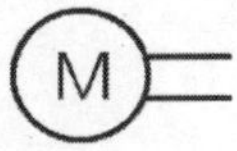	Electric motor
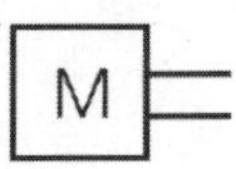	Internal combustion engine

end or double end rod. Besides to the possibility to show the cushioned design either from one side or from both sides.

Table 3.3: *Fluid power symbols (Linear and rotary actuators)*

Symbol	**Definition**
	Single direction, fixed displacement hydraulic motor
	Reversible, fixed displacement hydraulic motor
	Reversible, variable displacement hydraulic motor
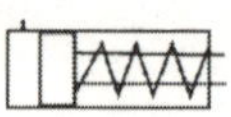	Single acting spring loaded actuator
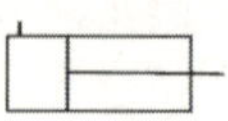	Simplified symbol of single acting actuator
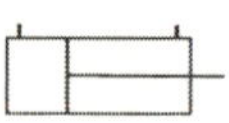	Double acting, single end rod actuator
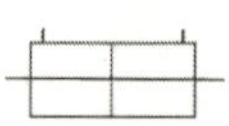	Double acting, double end rod actuator
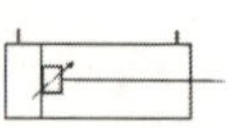	Double acting actuator with adjastable cushion

3.2.4 Symbols of Valves

Table 3.4 shows the different symbolic representations of valves. Different ways of control operations are shown with the corresponding definitions. The table explores how to represent the number of positions and number of ports (ways) for each control valve using boxes and arrows. As an example of how to define a control valve, look at Figure 3.10. The definition for this valve is:
Three position, four port, spring centered, solenoid controlled valve.

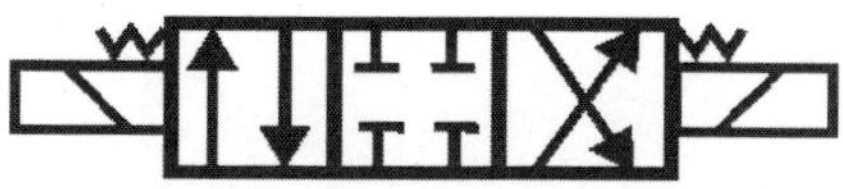

Figure 3.10: *Symbolic representation of a control valve*

3.2.5 Miscellaneous symbols

Table 3.5 includes other miscellaneous symbols of gas and spring loaded accumulators, presure and temperature indicators, besides to the symbols of heaters, coolers and filters. The FRL (Filter Regulator Lubricator) is shown in the table too. FRL is usually used in pneumatic systems.

3.2.6 comparison between Logic and fluid power symbols

Most of fluid power systems are controlled by electric or electronic components. These electronic components are represented by logic symbols. In some specific cases especially where there is high dust or moisture effect and there is a fear to get short circuits because of water and humidity, pneumatic controls are used instead of electric ones. Tables 3.6 and 3.7 show a comparison between logic symbols and fluid power symbols used for the same purpose.

Table 3.4: *Fluid power symbols (Valves)*

Symbol	**Definition**
	Butterfly manual ON-OFF valve
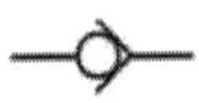	Non return (check) valve
	Pressure relief valve
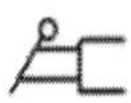	Manual hand control
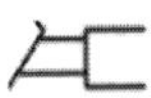	Pedal foot control
	Pressure pilot control
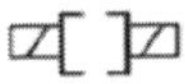	electric solenoid control
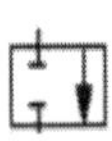	One position of a control valve
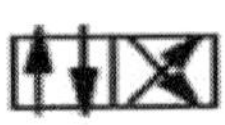	Two position four port valve
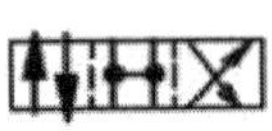	Three position four port valve

Table 3.5: *Fluid power symbols (Miscellaneous)*

Symbol	**Definition**
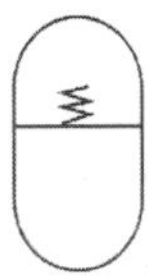	Spring loaded accumulator
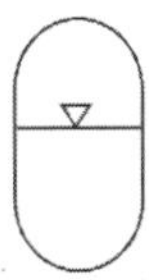	Gas loaded accumulator
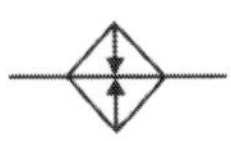	Heater
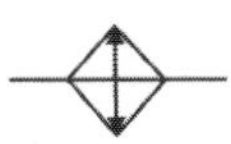	Cooler
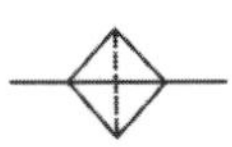	Filter or strainer
	Pressure indicator
	Temperature indicator
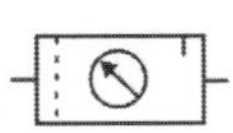	Filter Regulator Lubricator (FRL)

Table 3.6: *Comparison between logic symbols and fluid power symbols (basic symbols)*

Control element	Logic symbol	Fluid power symbol
And	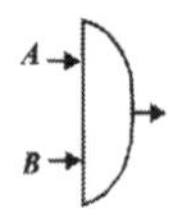	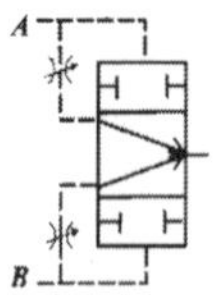
Yes	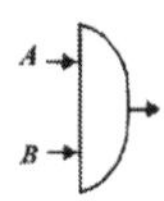	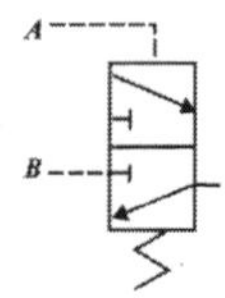
Or	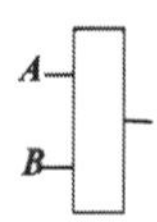	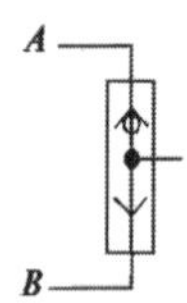
Not	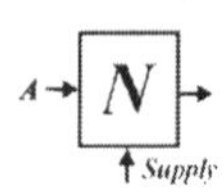	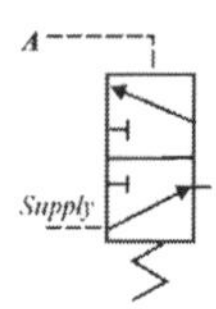
Nand	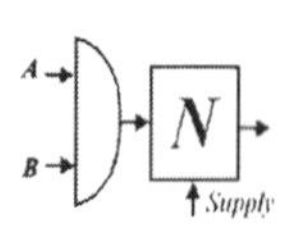	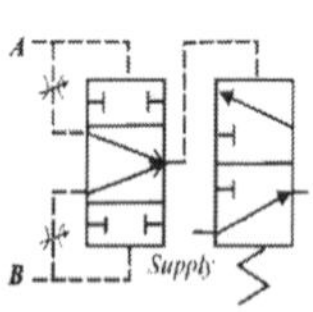
Nor	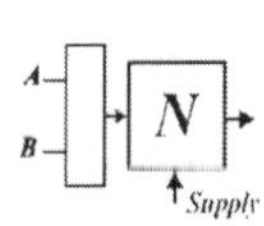	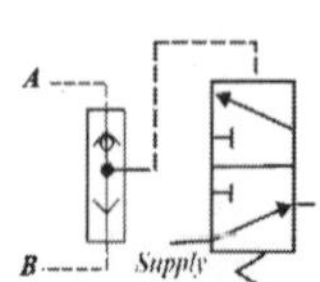

Table 3.7: *Comparison between logic symbols and fluid power symbols (miscellaneous symbols)*

Control element	**Logic symbol**	**Fluid power symbol**
Flip flop	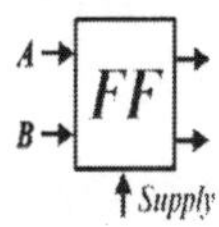	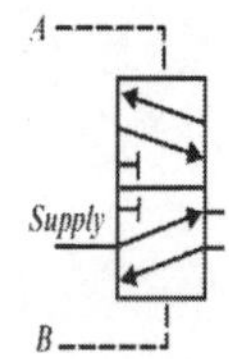
Memory	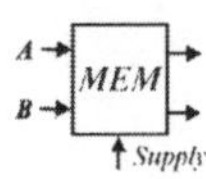	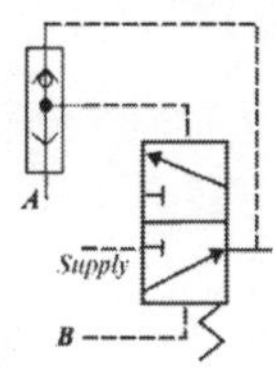
Step shot	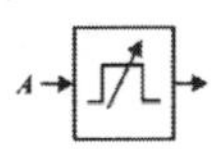	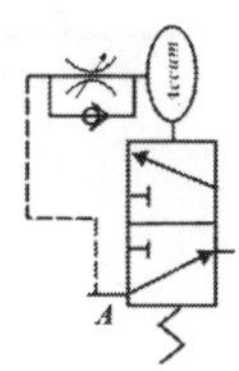
Delay timer	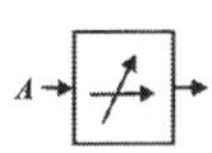	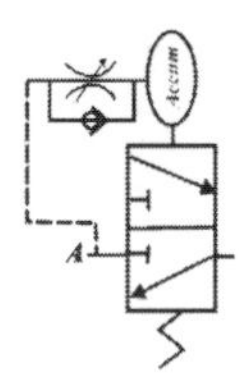

3.3 References

[1] F. Yeaple, *Fluid Power Design Handbook*, Third Ed., Marcel Dekker,1996.

[2] J. Pippenger and T. Hicks, *Industrial Hydraulics*, McGraw-Hill, 1979.

[3] H. Stewart, *Hydraulic and Pneumatic Power for Production*, Industrial Press Inc., 1977.

[4] M. Pinches and J. Ashby, *Power Hydraulics*, Prentice Hall, 1989.

Chapter 4

Electro-mechanical Controls

4.1 Electromagnetic background

When a bobbin φ is approached to a permanent magnet with the shown polarity $[S-N]$ (South-North), an electromagnetic force is induced in the bobbin and a current is induced if the ciruit is closed (see Figure 4.1).

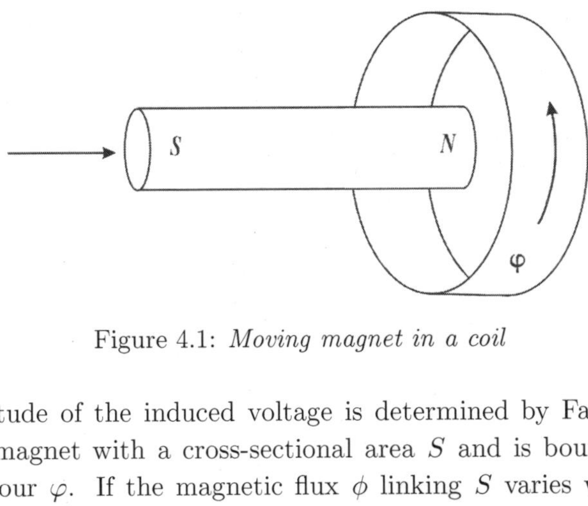

Figure 4.1: *Moving magnet in a coil*

The magnitude of the induced voltage is determined by Faraday's law. Assume a magnet with a cross-sectional area S and is bounded by the closed contour φ. If the magnetic flux ϕ linking S varies with time, a voltage V is induced around φ. This voltage is given by Faraday's law [1]:

$$V = -\frac{d\phi}{dt} = -L\frac{di}{dt} = -\frac{d}{dt}\int_S BdS \tag{4.1}$$

In circuit theory L is called the self-inductance of the element φ and V is called the voltage of the self-inductance. B is the magnetic field density [2]. The sense of direction of the induced voltage is determined by Lenz's law [3] which states that: *The voltage induced by a changing flux has a polarity such that the current established in a closed path gives rise to a flux which opposes the change in flux.* In other words, one can say that; in the induced current there is an electromagnetic inertia that opposes the variation of flux, like the reaction of mass inertia against the velocity variation. The voltage sense can be determined easily by applying the left hand rule, where, the pounce finger is the field and the index is the direction of movement, then the middle finger indicates the current. This polarity prediction assumes that the conductor experiences a magnetic force which opposes its motion. When a closed conducting loop is in motion with a relative velocity U with respect to a stationary magnet that generates a constant field B, this includes change in the shape. In this case Equation (4.1) can be applyied:

$$V = -\frac{d}{dt}\int_S BdS = \oint (U \times B)dl \tag{4.2}$$

If the velocity U and the field B are at right angles and the conductor is normal to both, then a conductor with a length l will have a voltage:

$$V = BlU \tag{4.3}$$

According to the previous assumptions, B and l are constants. This leads to express the linear relation between the voltage and the velocity as follows:

$$V = GU \tag{4.4}$$

Where G is the *transduction constant* (in [V/m/s] or [N/A]) which can be called also the *Lorentz force constant*. From the previous discussion, one can see that the relation between the magnet velocity and the induced voltage depends strongly on the magnetic field density B and the length of the conductor l. In the case of coaxial cylindrical magnet and cylindrical coil, the length of the conductor is equal to the length of one turn of the coil multiplied by the number of turns.

4.1.1 Forces in the magnetic field

If a charged particle Q is moving in an external magnetic field B with a velocity V,as shown in Figure 4.2, the force F exerted on this particle is given by

$$F = Q(V \times B) \tag{4.5}$$

The direction of this force is defined by the right hand rule. In Figure 4.2, if the magnetic field B goes into the page and the particle moves to the right, then the force will be upwards.

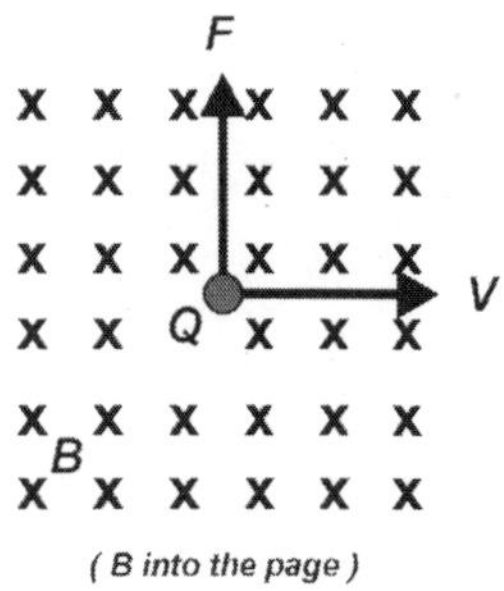

Figure 4.2: *Force on a charged particle in a magnetic field*

Since the rate of change of the charge with respect to time is the current,

$$I = \frac{dQ}{dt} \tag{4.6}$$

Replacing the charged particle by a current carrying conductor with a length l, the differential force on the conductor becomes

$$dF = dQ(V \times B) = (Idt)(V \times B) = I(dl \times B)$$

Where $dl = Vdt$. If the conductor is straight and the field is constant along it, both sides of the differential equation can be integrated to get

$$F = I(L \times B) = ILBSin\theta \tag{4.7}$$

Again, the direction of the exerted force is defined by the right hand rule. One can imagine, as in Figure 4.3, the field lines as stretched rubber bands that push the conductor exerting this magnetic force (the imagination of Michael Faraday). In the case when the conductor is a circular bobbin with a radius R, and N number of turns, equation (4.7) takes the form

$$F = ILBN \tag{4.8}$$

Such that $L = 2\pi R$. Here, $Sin\theta = 1$ because the magnetic field is always normal to the current.

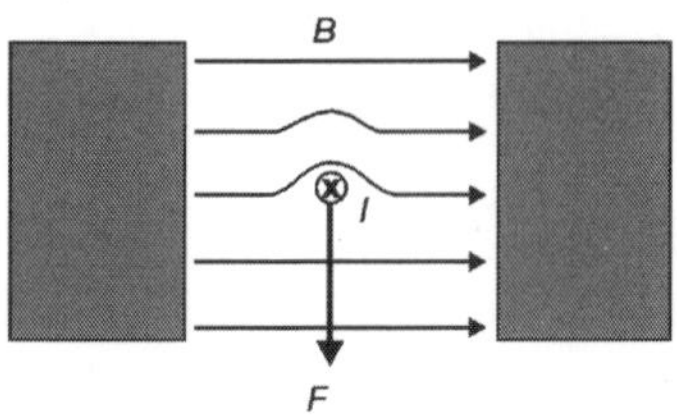

Figure 4.3: *Force on a conductor due to magnetic field (Faraday's imagination)*

4.2 Solenoids and relays

Solenoid is an electromechanical device consists of a cylindrical coil wrapped around a coaxial armature (see Figure 4.4). The magnetic armature is free to move axially inside the coil. The solenoid is an actuation device used to convert electromagnetic energy into linear motion. When an electric current passes in the coil, the induced electromagnetic force in the magnetic field forces the magnetic armature to move axially back or forth according to the current's sense of direction applying the right hand rule. Solenoids are used to move the spool of the spool type valve in both, hydraulic and pneumatic valves forming the so called *Solenoid Valves.* Spools are pushed by the solenoid armatures to open or close ports and change position of valve. Figure 4.5 shows a real solenoid valve used for pneumatic applications.

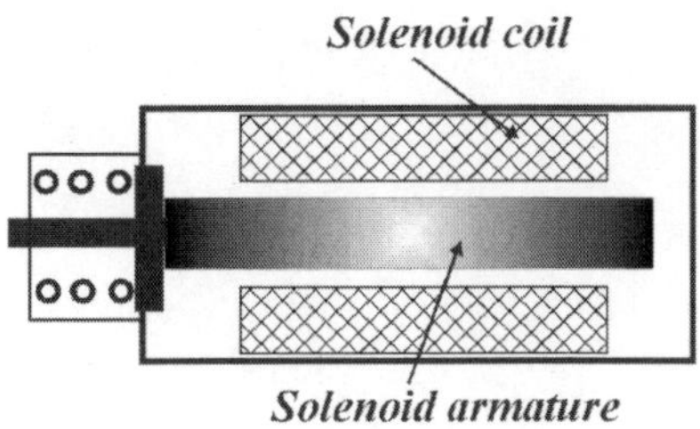

Figure 4.4: *Electromechanical solenoid design*

Figure 4.5: *Pneumatic solenoid valve (Courtesy of Fenghua Xingyu)*

Another main application on using electromagnetic power to control fluid power systems is *Relays.* Relay is an electromagnetic switch that consists of a contacting arm pulled and pushed by a solenoid (see Figure 4.6). When passing a low voltage electric current in the coil of the solenoid the induced electromagnetic force moves the armature back or forth causing the contacting arm to move and close the high voltage circuit to operate the required function. De-energizing the coil of the relay returns the arm to its center point opening the high voltage circuit and stopping the operation. Since relays are switches, a relay will switch one or more poles contacting when energizing the coil in one of three ways:

- Normally open **(NO):** The circuit is normally disconnected when the relay is de-energized, it contacts when the coil is energized.

- Normally closed **(NC):** The circuit is normally connected when the relay is de-energized, it disconnects when the coil is energized.

- Change over **(CO):** Contacts control two circuits: one normally-open contact and one normally-closed contact with a common terminal.

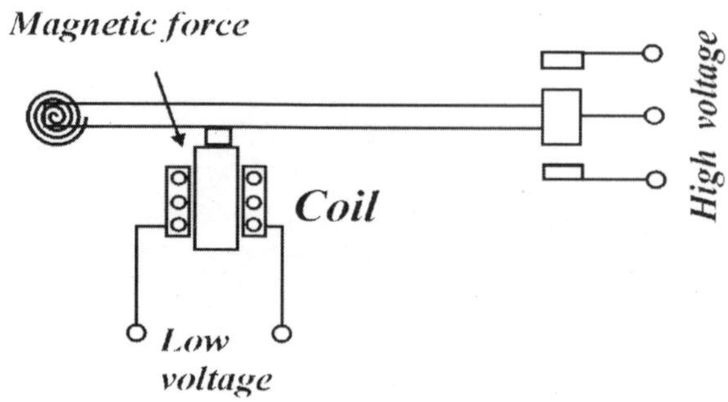

Figure 4.6: *Electromagnetic Relay*

4.3 Voice coil linear actuator

Another type of linear actuator is the so called *Voice Coil Actuator.* Voice coil actuator depends mainly on using magnetic field to induce an electromagnetic force. On the contrary of the solenoid, the moving part in the voice coil is the coil itself. Another advantage of the voice coil is that it can be used for precise displacements but for short strokes.

4.3.1 Basic Architecture

Figure 4.7 shows a cross-section of the internal basic design of a voice coil actuator. This type of actuators is made up of two components; a moving part and a fixed part. The upper part in the figure is the moving part and consists of a group of wires wound in a tubular form around the

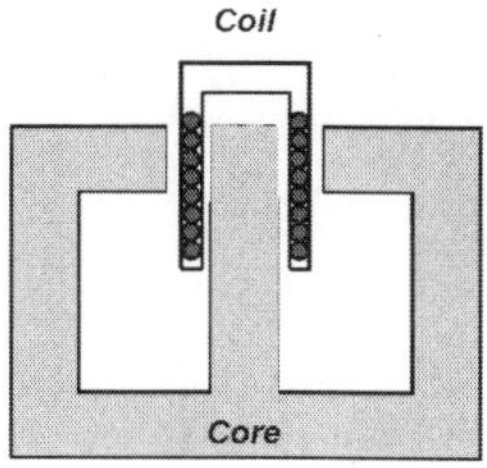

Figure 4.7: *Basic schematic architecture for a voice coil*

coil holder. The holder can be manufactured out of plastic or fibre or any other nonconductive material depending on the load carried by this part to reduce the effect of eddy currents on the motion. The coil itself is formed out of a conducting wire like copper. The diameter of the wire and the dimensions of the holder are determined according to the application. The stationary member is made up of a permanent magnet. The polarity of the magnet can be radial or axial according to the design, and this is one of the basic concepts and differences between the three designs that will be explained later. This member can be not only a permanent magnet but it can also contain a ferromagnetic material like the iron or some steel alloys to pass the magnetic flux generated by the permanent magnet. In the voice coil, the magnetic cycle should be closed to have a continuous flux. Although there is an air gap that allows the coil to pass through, but this air gap should be as small as possible to reduce the fringe in the flux at that point [4].

As mentioned in the previous section, when a current carrying conductor crosses the stream lines of a magnetic field; a force is exerted on the conductor. This force is called *Lorentz force* and is a function of the magnetic field density B, the current applied in the conductor I, the length of the conductor and the angle between the magnetic field and the current. The force is maximum when the field is normal to the current. It creates a linear motion in the moving part because the force is linearly proportional to the current applied in the coil, and this leads to achieve a good control. The displacement (stroke) moved by the voice coil can vary from microns up to a few millimetres. From the force point of view, the voice coil is considered a weak force producer relative to its size, in other words, it is not as weight-power-efficient as other devices.

4.3.2 Modelling of the voice coil

The relative linear motion between the two parts of the voice coil is mainly caused by the current passing across the magnetic field. The output force of the actuator is linearly proportional to the input current. The ratio between the force and the current is called the actuator constant, this ratio is constant for each actuator and depends on the size, design and shape of the actuator itself.

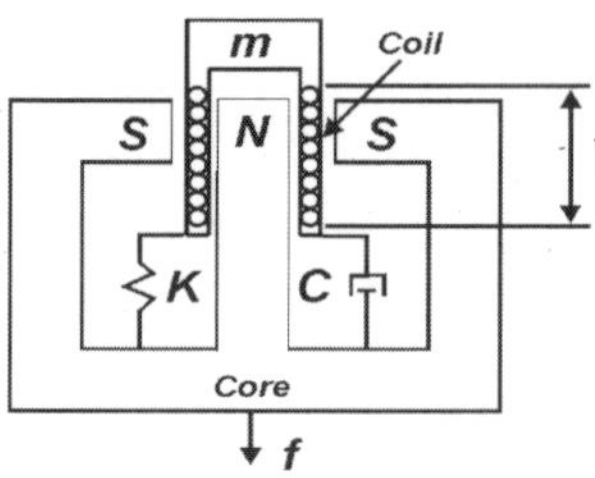

Figure 4.8: *installation of a voice coil actuator against a fixed reference*

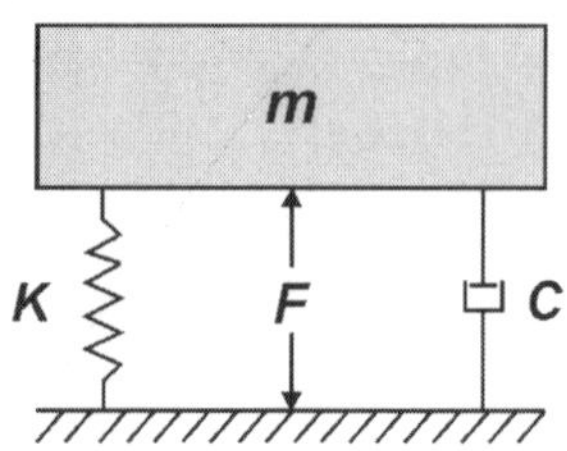

Figure 4.9: *Mass, spring and dashpot representation of the voice coil*

To study the behaviour of a voice coil, it can be installed against a fixed reference. In this case, the mass m will act as a charged load, (Figure 4.8). The actuator is mounted on the top of a force sensor to measure the output force. This installation can be represented by a simple mass with a spring and a dash-pot, as shown in Figure 4.9, Which leads to a single degree of freedom model with the following equation of motion:

$$m\ddot{x} + C\dot{x} + Kx = F \tag{4.9}$$

But $F = GI$, where G is the *Lorentz force constant.* Applying the Laplace transform on Equation (4.9) we get:

$$s^2x + 2\xi\omega_n sx + \omega_n^2 x = \frac{G}{m}I$$

$$\frac{x}{I} = \frac{\frac{G}{m}}{s^2 + 2\xi\omega_n s + \omega_n^2} \tag{4.10}$$

$$\frac{F}{I} = \frac{-Gs^2}{s^2 + 2\xi\omega_n s + \omega_n^2} \tag{4.11}$$

Where the natural frequency $\omega_n^2 = \frac{K}{m}$ and the damping part of the equation is $2\xi\omega_n = \frac{C}{m}$, including contributions from the structure and from the back electro-motive force ($e.m.f$), because here the $e.m.f$ appears as an additional damping in the system, [5]. The transfer function expressed in Equation (4.11) is plotted in Figure 4.10. This Bode plot enables us to predict the characteristics of the system.

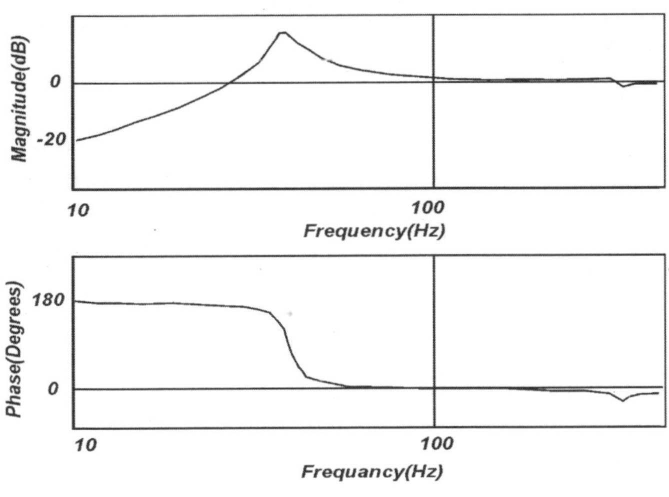

Figure 4.10: *Transfer function of the voice coil with the current as an input and the force as an output*

To determine the force constant G of the actuator which leads to define the force, a blocked transfer function is measured (see Figure 4.12). To obtain this plot one can block both sides of the voice coil actuator to prevent it from moving and apply a current into the coil as shown in Figure 4.11.

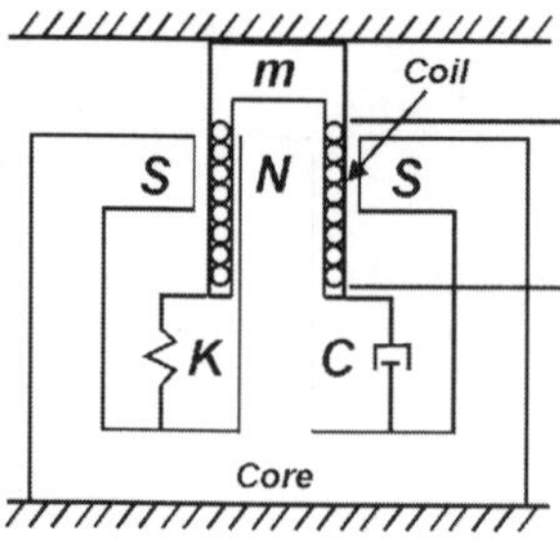

Figure 4.11: *Blocked installation of a voice coil*

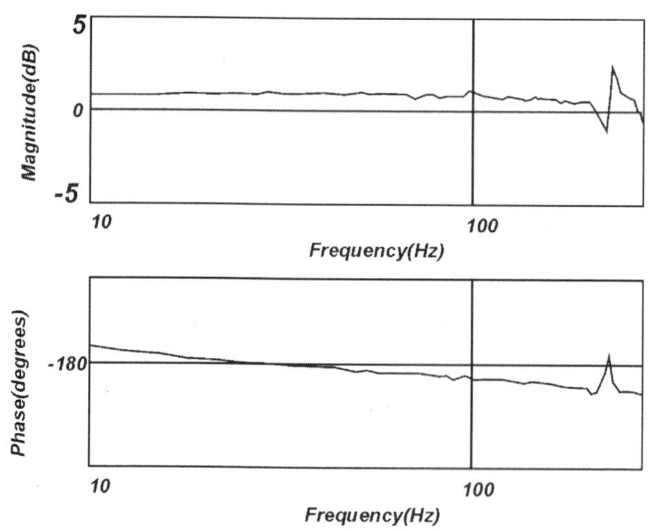

Figure 4.12: *Blocked transfer function of the voice coil with the current as an input and the force as an output*

4.3.3 Configurations of voice coil actuators

Several designs can be proposed to build an actuator based on the voice coil principle. Here, three configurations have been selected considering that they are the most common used designs.

Radial toroid voice coil actuator

The first design depends on employing a toroidal permanent magnet with radial polarity. The magnet is included coaxially in a cylindrical ferromagnetic material. A core from the same material exists along the central axis of the cylinder (Figure 4.13). Thanks to the large axial dimension of the permanent magnet, relatively large number of turns in the coil can be wound, which enables to obtain higher force. One of the restrictions that may face the designer here is the difficulty to obtain a large axial length for one magnet, but still there is a possibility to stack several magnets on top of each others. Another difficulty is that it is more expensive to produce a permanent magnet with radial polarity which increases the cost of the actuator. In order to minimize the volume and weight of the actuator, a special steel alloy with high magnetic permeability is used. Special alloys are used to avoid reaching the saturation limit in the ferromagnetic part. All the coil turns are accounted here in the force calculation. Because the flux generating area of the magnet is larger than the area of the bobbin coil [4].

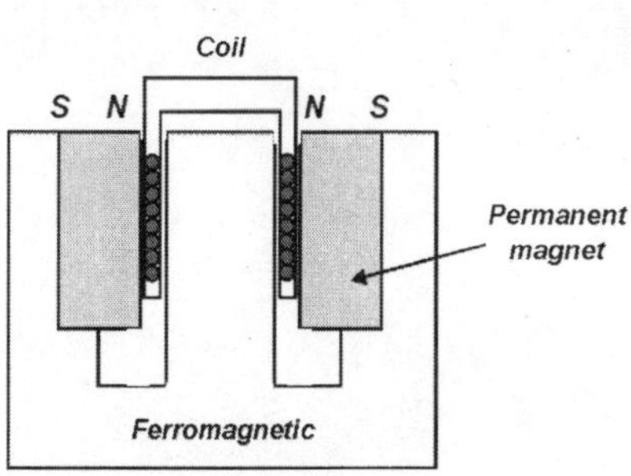

Figure 4.13: *Radial toroid voice coil actuator*

Axial toroid voice coil actuator

Another way of the design is shown in Figure 4.14. In this type, a toroidal shaped magnet is used here too, but this time the polarity is selected to be axial. To guide the magnetic flux generated by the magnet, a hollow disk of ferromagnetic material is mounted on the top of the magnet. This piece of metal aims at concentrating the magnetic field at the air gap where it crosses the coil current. On the other side of the magnet, another solid disk is mounted. In the center of this disk and along the central axis, a core of the ferromagnetic material is installed. This type of configuration is easier to manufacture and is more economic than the previous one. To concentrate the flux in the air gap, the axial area near the gap should be minimized. This leads the designer to try to increase the number of turns of the coil to keep the force level, which, unfortunately, needs to increase the air gap. To do so, an optimization technique should be held to compromise all the dimensions including the air gap. In this configuration the axial distance of the coil is larger than the thickness of the hollow disk to enable the magnetic flux to cross the same number of turns during the stroke of the motion.

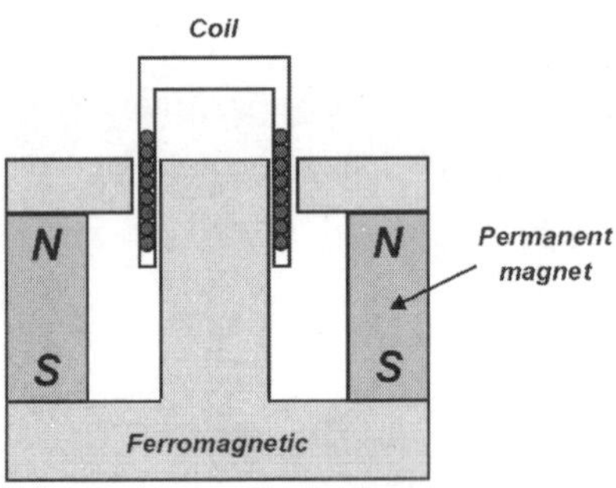

Figure 4.14: *Axial toroid voice coil actuator*

Axial disk voice coil actuator

This configuration is different from the previous two by the shape of the magnet and the situation where it is installed. Here the magnet is a circular disk put just under the core along the central axis of the cylinder (see Figure 4.15). The magnetic flux, flowing axially out of the permanent magnet, passes through the ferromagnetic material to be concentrated at the air gap where the current carrying coil crosses the field lines generating a force. The central core should be well centred over the magnet and inside the outer cup to keep a uniform air gap and avoid flux leakage. The axial distance of the coil here also should be larger than the thickness of the ferromagnetic material at the air gap to cover the same number of turns during the motion of the bobbin [6].

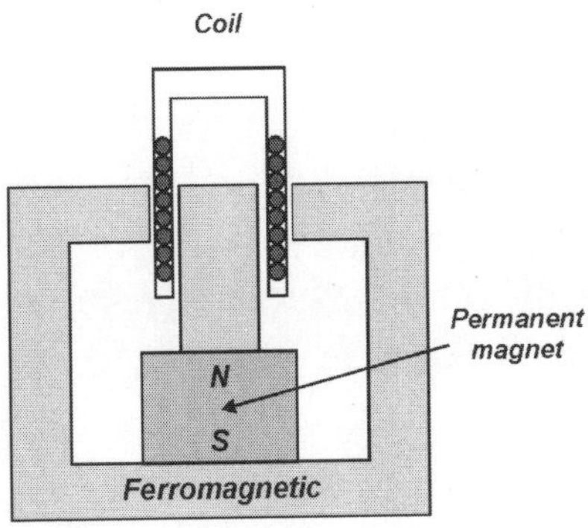

Figure 4.15: *Axial disk voice coil actuator*

4.4 Servo valves

Single stage spool valve (shown in Figure 4.16) is the basic design of the servo valve where the spool is driven by the torque motor via a mechanical mechanism to determine the final position. The main problem here is the static friction between the spool lands and the casing. This problem is solved by adding a dither signal to the main control signal. The dither signal is a signal with a very small amplitude and high frequency that reaches to 100 Hz. This signal oscillates the spool around its rest position with very small strokes to overcome the static friction force and keep the spool in continuous dynamic motion.

Another type of servo valves is the flapper type shown in Figure 4.17. The

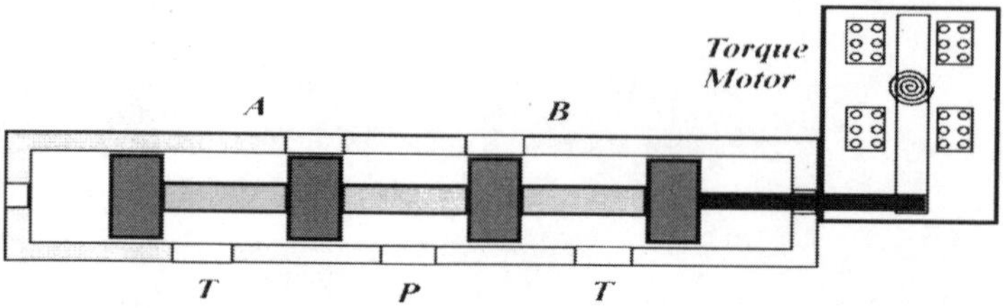

Figure 4.16: *Single stage spool valve directly operated by a torque motor*

principle of this type is based on moving the flapper plate to the right or the left by the magnetic force induced by the electromagnetic coils. When the flapper moves to the left it closes the outlet of the left nozzle increasing the pressure in that side and causing the spool to move to the right. The value of the produced force here is controlled by the restrictor mounted on that pipe.

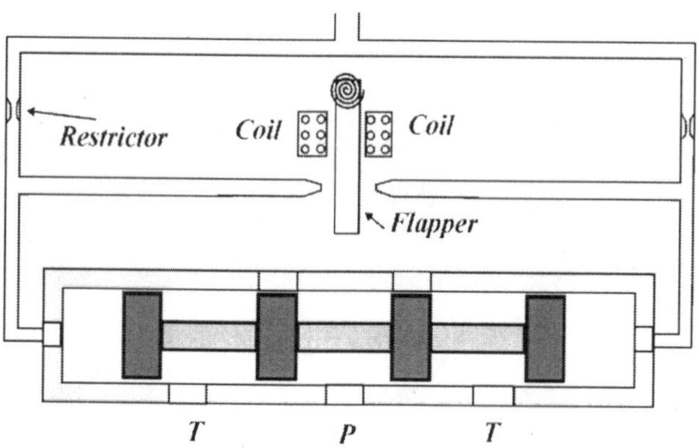

Figure 4.17: *Flapper servo valve*

The third type of servo valves is the jet servo valve shown in Figure 4.18. The main jet pipe is pushed right or left here by means of the electromagnetic force induced by the coils. Pushing the jet to the right causes the fluid to flow from the jet to the right side pipe forcing the spool to move from the right to the left.

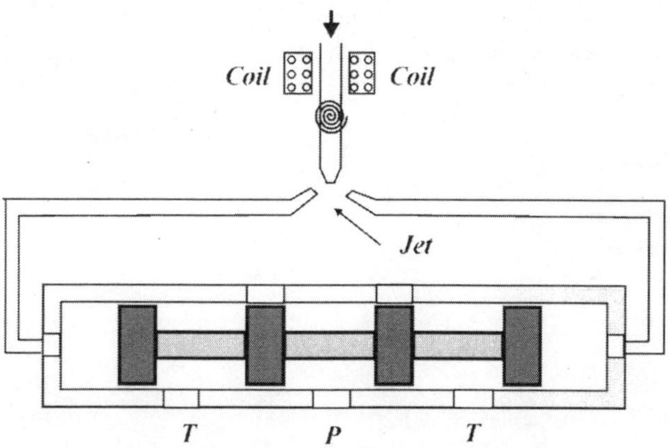

Figure 4.18: *Jet servo valve*

4.5 Proportional valves

In servo valves, the armature moves in a full stroke either open or close. This can be applied for small size applications but in most of the heavy duty applications there is a need to open or close the valve gradually to avoid shocks and rapid transition response from one side and to control flow and pressure according to the load from the other side. This gradual movement can be achieved by using proportional solenoids or voice coil actuators. The force exerted by the armature of the solenoid here is proportional to the current input to the coil of the solenoid. While the relation between the armature force and the displacement of the armature is constant as far as the armature is immersed in the magnetic field of the coil where the force begins decaying after that. Voltage is not adequate for solenoid control because the coil resistance is temperature dependent. A current amplifier is needed here to provide the circuit with an amplified and conditioned current output to control the solenoid as shown in Figure 4.19. The current feedback loop is necessary to reduce the hysteresis caused by the change of the current direction. The two horizontal lines above and below the valve indicate the possibility of infinite positioning of the spool.

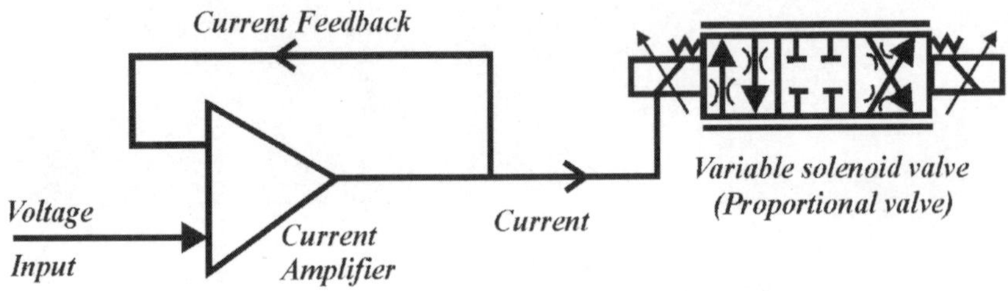

Figure 4.19: *Current control of a proportional solenoid valve*

The block diagram and time response of the current amplifier required to control the proportional valves is shown in Figure 4.20. The amplifier consists of a ramp up step generator that determines the rate of acceleration of the spool to reach the constant speed of motion at the maximum current and a ramp down signal generator that determines the decceleration at which the spool moves before reaching the final position. Another component in the amplfication system is the dither. A dither signal is a low level signal superimposed to the input; it is an AC oscillating signal at a rate of about 100 Hz. This signal keeps the spool oscillating around its rest position to overcome static friction and increase the spool's response. The feedback signal fed to the amplifier can come from the following sources:

- *Hysteresis control:* the output current of the amplifier is measured and fed back to the amplifier again to reduce the hysteresis effect which is considered very much higher in proportional valves than servo valves.

- *Spool position control:* a displacement transducer can be attached to the spool of the valve measuring its position and feeding it back to the current amplifier to be corrected according to the required position.

- *Load speed or position control:* the position or the speed of the output load can be measured and fed back to the amplifier to be determined as desired by the operator.

Although there is a feedback loop in the hysteresis control and spool position control but still the circuit is considered an open-loop unless there is a feedback signal coming from the load to control the output.

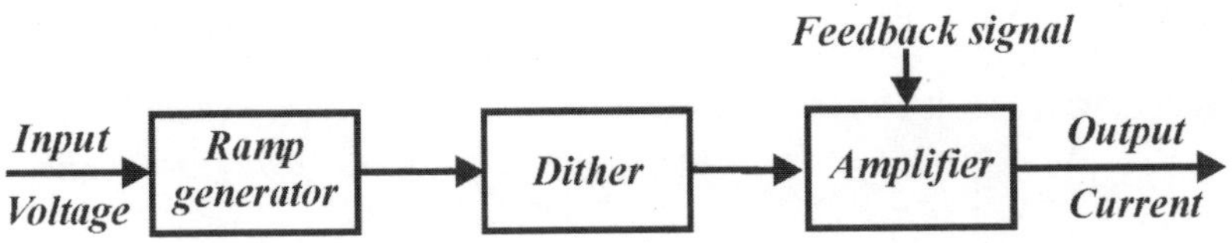

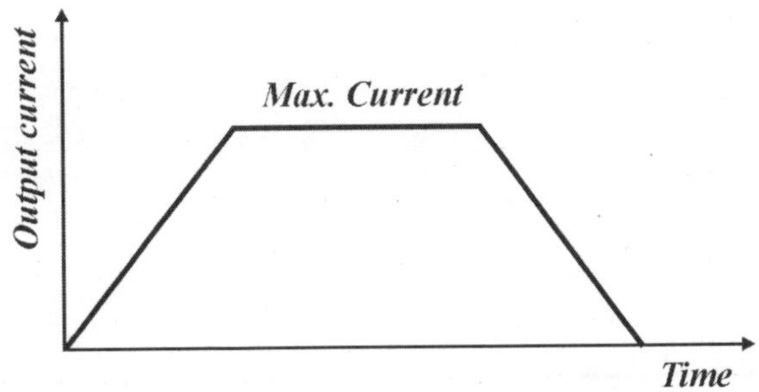

Figure 4.20: *Block diagram and time response of the amplifier used for proportional valves*

Although servo valves are more accurate and convinient than proportional valves for their better response and hysteresis effect, but they are much more expensive to manufacture. Thus, proportional valves are considered an economic and achievable solution when high accuracy and fast response are not essential. Table 4.1 shows a comparison between servo and proportional valves.

Variable current is used to control the force of the proportional valves. Using voltage control is not precise enough because the increase in temperature changes the resistance of the coil. Therefore, the output force is proportional to the input current. A control spring can be added to the valve to help in spool stability and to have more accurate control although this means the need for a higher force to overcome the spring stiffness.

Table 4.1: *Fluid power symbols (Miscellaneous)*

Characteristic	**Proportional valve**	**Servo valve**
Valve lap	Overlap with dead zone	Zero or underlap without dead zone
Response time	40 - 60 ms	5 - 10 ms
Operating frequency	10 Hz	100 Hz
Hysteresis	1% - 5%	0.1%

As mentioned before, it is difficult and expensive to manufacture zero-lap lands spool valves. Thus, over-lap spool valves are used in proportional valves which implies having a dead zone in which there is a current value (200 mA) without having any fluid flow from the ports as shown in Figure 4.21. To avoid the problem of the dead zone and improve the valve response, notches are added to the lands of the spool as shown in Figure 4.22.

4.6 Spool position control

The flow rate through the spool valve can be controlled precisely by controlling the position of the spool that allows to open or close gradually the operation ports of the valve. Current feedback control can be done for the spool postion control as shown in Figure 4.23. A Linear Variable Differential Transformer (LVDT) or a potentiometer can be used as a transducer fixed to the spool of the valve. The output voltage of the transducer is

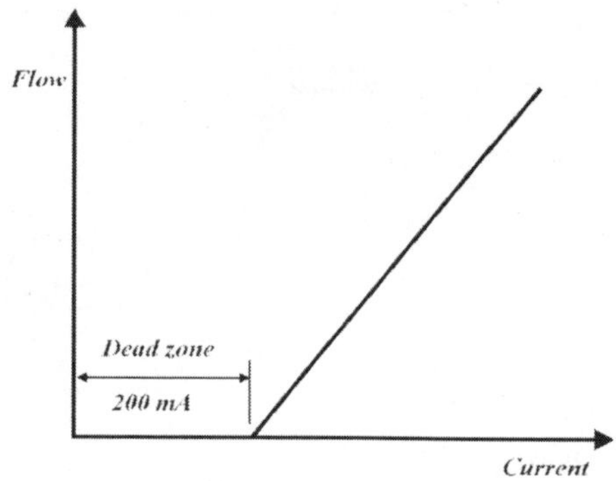

Figure 4.21: *Flow current relationship in proportional valves*

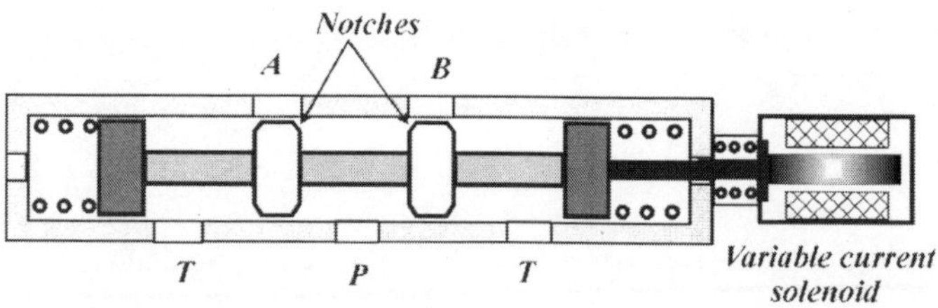

Figure 4.22: *Proportional directional control valve with notched spool*

proportional to the displacement of the spool, this position is fed back to the current amplifier to be compared with the required input position previousely determined by the operator. According to this comparison, the current fed into the solenoid is adjusted to fit the required position. Figure 4.23 shows an example of spool position feedback control for a two directional hydraulic motor, where a transducer is used to control the position of the spool but this is not enough to control the actual flow reaching to the load. Therefore, a closed-loop speed feedback control is applied by measuring the speed of the hydraulic motor (using a tachogenerator) and feeding it back to the acurrent amplifier to be compared to the desired input speed. The influence of the dead zone in the proportional valve should be taken into account in spool position control.

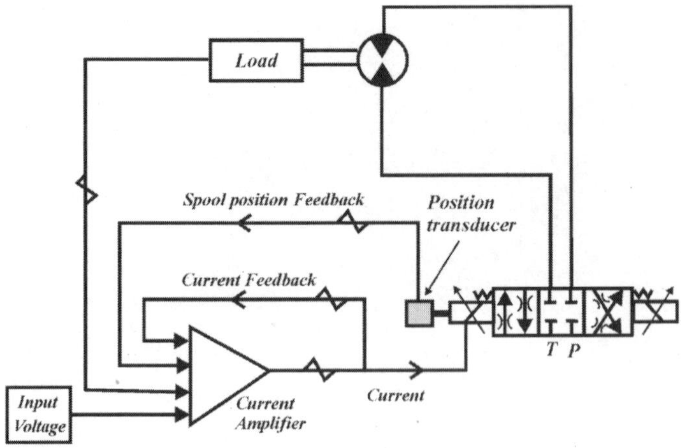

Figure 4.23: *Closed-loop speed control with spool position control for a hydraulic motor*

4.7 Pressure control

In single stage proportional relief valve shown in Figure 4.24, the solenoid acts on the poppet of the valve with a force proportional to the applied current. When the valve is Normally Open (NO) by means of the spring inside the solenoid, the solenoid is actuated to close the valve, the fluid pressure needs to overcome the difference between solenoid force and the spring force to open the valve again. In the other case, the valve is Normally Closed (NC) and the solenoid is used to change the stiffness of the spring changing the force needed to open it and thus changing the fluid pressure required to open the port and allow the fluid to pass.

Increasing the control current flowing into the solenoid increases the pressure needed to overcome the solnoid force and to open the nozzle of the valve having a maximum flow through it. This relation is shown in Figure 4.25 where it shows that the relation is not really linear but there is some nonlinearity in the shown curves.

Another main example of using solenoids to control the fluid pressure is shown in Figure 4.26. This system shows a conventional pressure regulating

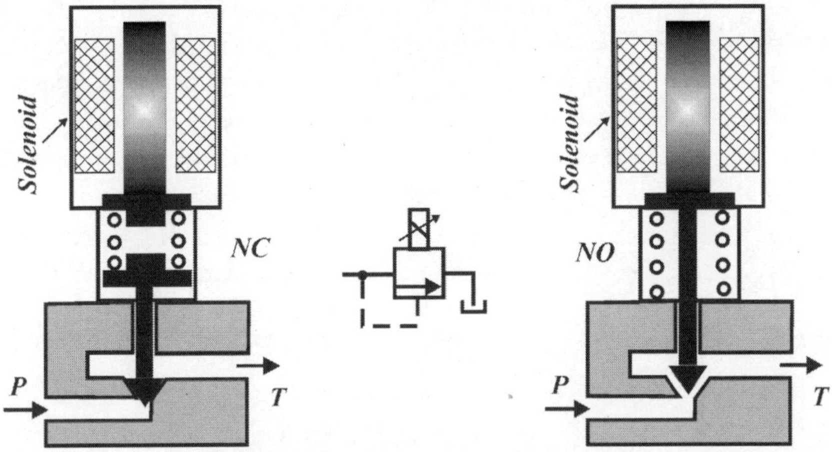

Figure 4.24: *Pressure relief valve with solenoid control; Normally Open (NO) and Normally Closed (NC)*

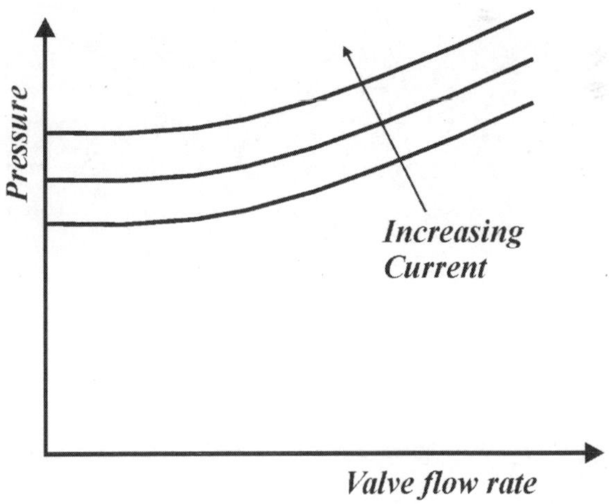

Figure 4.25: *Relationship between pressure and flow rate of fluid through the valve with increasing the control current of the solenoid for NC valve*

valve. When the solenoid is energized, the fluid flows through the spool of the valve with full pressure value. When the solenoid is de-energized, the spring returns the spool back to its rest position where part of the flowing fluid is fed back to the tank. Intermediate positions of pressure reduction by returning part of the fluid back to the tank can be obtained by changing the control current flowing to the solenoid. Eventually, the pressure of the output fluid is proportional to the control current of the solenoid.

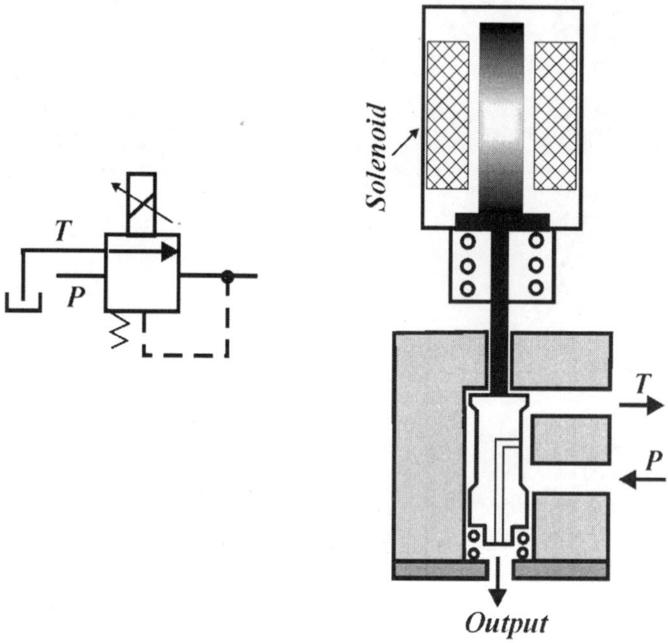

Figure 4.26: *Pressure reduction valve*

Single stage proportional valves are limited to low flow capacities, less than $5l/min$. To obtain higher flow rates, two stage proportional directional control valves are required like the one shown in Figure 4.27. When the left solenoid is energized, it opens the pressure relief valve allowing the pilot fluid to flow to the left side chamber of the directional control valve. Pressure accumulates behind the spool till it overcomes the spring force acting on the spool. Hence, the spool moves to the right opening the way to the operational pressure to flow through port B. This spool is overlapped with notches and thus it opens gradually having a dead zone of 40-60 ms. Flow through the valve increases nonlinearly with the increase of the control current but this changes as the pressure drop across the valve increases as shown in Figure 4.28.

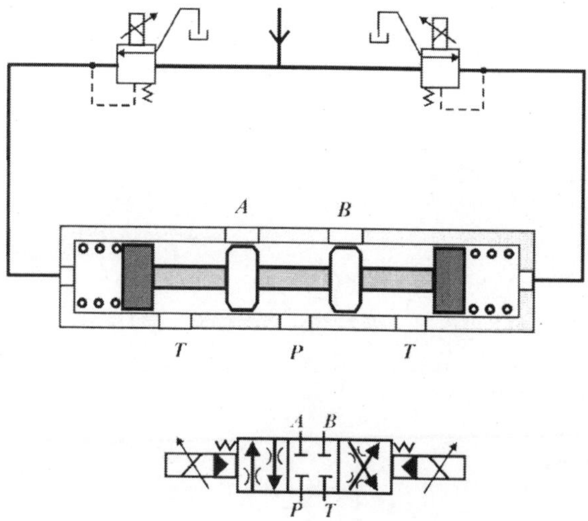

Figure 4.27: *Two stage proportional directional control valve*

4.8 Douple flow control

Flow in proportional valves is usually controlled by adjusting the spool position with changing the current of the solenoid but this flow is limited to the size of the port. If this flow is not enough, two ports are connected together to double the flow as shown in Figure 4.29. In this case too, the flow is proportional to the current passing through the solenoid but the relationship here is not linear and can be adjusted by adding notches to the lands of the spool.

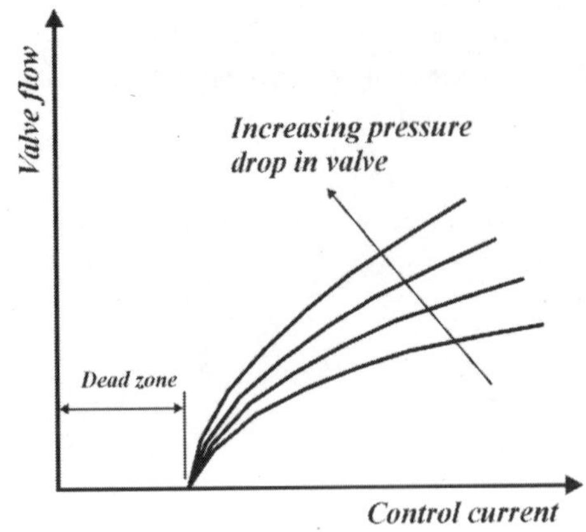

Figure 4.28: *Flow-current relationship with increasing pressure drop in two-stage proportional valve*

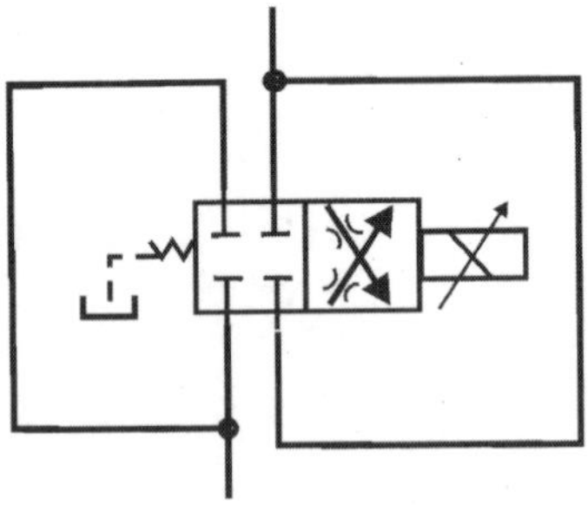

Figure 4.29: *Double flow of four port directional control valve*

Combining the ports to get a double flow can be accompanied by a pressure drop downstream or upstream. Doubling the flow can be obtained without being influenced by the pressure drop by adding a pressure compensation setup that can compensate for the pressure loss by feeding a pilot line of the pressure back to the pressure relief valve and varying the flow orifice to maintain a constant pressure drop as shown in Figure 4.30.

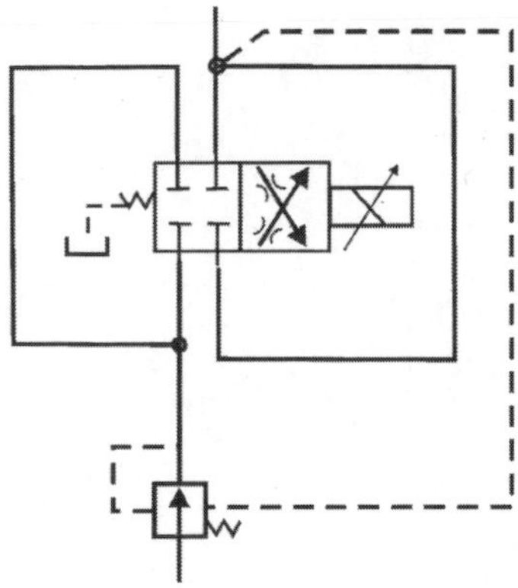

Figure 4.30: *Pressure compensation in double flow of four port directional control valve*

4.9 Control of actuators

The aim for controlling actuators is to regulate the speed of the ram to fit for specific applications. Meter-in and meter-out is considered the main technique used to control the speed of the different actuators as shown in Figure 4.31. This technique is based on using non return (check) valves to direct the flow to a by-pass line and force the fluid to pass through a flow control valve that changes the flow rate which varies the speed of the piston. changing the sense of direction of the check valves determines if the control is meter-in or meter-out. In the case of meter-in control, the flow of fluid entering the piston is controlled while in the case of meter-out control, the flow control valve regulates the fluid leaving the piston.

Another technique used to control the piston speed is by using a cam connected to the piston rod as shown in Figure 4.32. The cam is designed in a way to be able to push the spool of the two position control valve leading the fluid to pass through a flow control that in turn regulates the speed of the piston.

The same techniques used to control the speed of linear actuators can be used to control the speed of the rotary hydraulic motors. The disadvantage

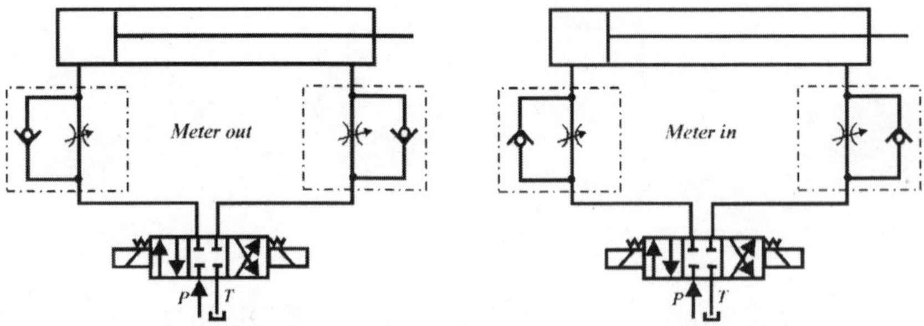

Figure 4.31: *Meter-in and meter-out actuator speed control*

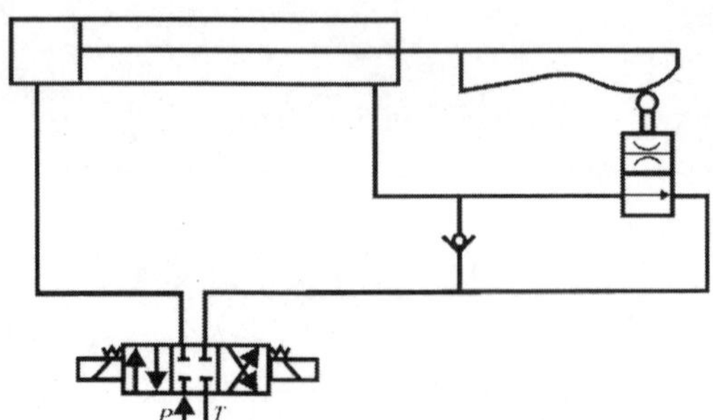

Figure 4.32: *Cam operated speed control of the actuator*

of these techniques is the slow response due to using proportional valves. In specific applications where rapid response is a need, servo controls are used. Variable displacement pump servo control technique is considered the best way to control hydraulic motors.

The position of the linear actuators (pistons) can be controlled by adding position transducers to the spool of the directional control valve besides to using position sensors on the load to overcome the dead zone difference and feed a signal back to the solenoid through the current amplifier as discussed previousely.

4.10 Control of pumps

Pressure and flow are two variables required to be controlled in pumps. The simplest way used to control the pressure and have a constant pressure value in the system is using a proportional pressure relief valve (PRV) that can be adjusted to the required pressure. The pressure relief valve will open only when the pressure in the system exceeds the requied valuc where it allow the fluid to pass returning back to the tank. Figure 4.33 shows a control circuit used to control the pressure and the flow of a variable displacement pump. The flow of the pump returns back after a specific pressure to the tank via the pressure relief valve D untill applying current to the solenoid of the directional valve A. Moving the spool of valve A to the right allows to control the flow by the throttle in valve A. Leaving the pressure relief valve B closed increases the pressure in the pilot line and moves the spool of the three position three port directional control valve C to the right increasing the displacement of the pump. If the system pressure is larger than the pressure value set to valve B, this valve opens passing the fluid to the tank and the spool of valve C moves to the left opening the large pump piston to the tank.

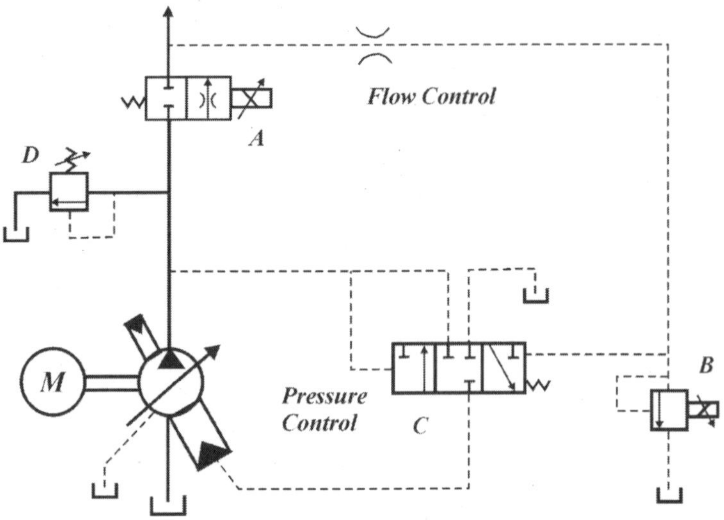

Figure 4.33: *Pump proportional control, pressure and flow*

4.11 References

[1] J. D. Kraus, *Electromagnetics*, 4th ed., McGraw-Hill international editions, electrical engineering series, 1992.

[2] J.A Edminister, *Schaum's theory and problems, "Electromagnetics".* Schaum's outline series, McGraw-Hill, Inc., 1993.

[3] A. Fouille, *Electrotechnique a l'usage des ingenieurs.* Dunod, Paris, 1961.

[4] Kimco magnetic division "BEI", *Voice coil actuators and applications guide.* BEI technologies Inc., 1998.

[5] A. Preumont, *Vibration control of active structures*, 1st. ed., Press Universitaire de Bruxelles, pp33-34, 1996-1997.

[6] Brüel & Kjaer, *Brüel & Kjaer, catalog of shakers,* Brüel & Kjaer ,1992.

Chapter 5

Hydraulic Circuits

5.1 Introduction

This chapter will focus on the design and analysis of different hydraulic circuits and their applications. The circuits discussed here are the systems controlled by hydro-mechanical and electro-mechanical controls only. Any mechanically controlled hydraulic circuit consists mainly of:

- Hydraulic tank (reservoir filled with hydraulic liquid).
- Hydraulic pump (reciprocating, gear, vane or any other type).
- Hydraulic actuator (linear piston or rotary motor).
- Hydraulic valve (pressure relief, flow control or directional control).
- Prime mover (electric or internal combustion motor)

Several considerations must be taken into account when designing a hydraulic system:

- Saftey of operator and operation.
- Efficiency and performance of the whole system.
- Cost requirements.
- Simplicity and easiness.

The following sections will discuss different basic hydraulic circuits using the fluid power symbols listed in the previous chapter that conform with the Amercian National Standards Institute (ANSI).

5.2 Single-acting hydraulic cylinder

5.2.1 Application

The simplest circuit is the single-acting spring-loaded cylinder shown in Figure 5.1. It is used when the needed action is one sense of direction like pushers and ejectors in production lines.

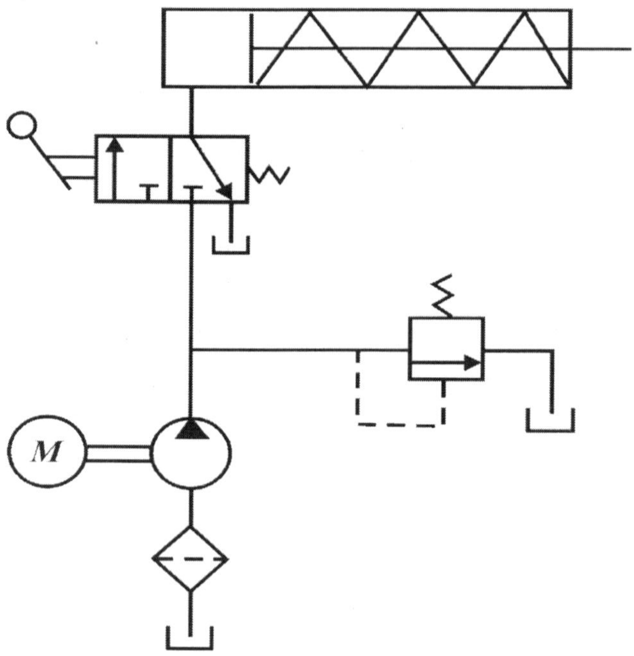

Figure 5.1: *Hydraulic single-acting cylinder*

5.2.2 Analysis

1. The whole circuit is supplied by a single directional constant displacement pump that pumps the fluid from a tank through a filter or strainer.

2. The pump is followed by a pressure relief valve calibrated to allow

the fluid to run away from the circuit back to the tank when the pressure inside the circuit reaches the calibrated pressure value.

3. This circuit is controlled easily by the two-position three-port manually-controlled directional valve shown.

4. The valve is forced normally by a spring to the right position that draws the fluid from the piston back to the tank and the spring inside the piston forces the rod to return back (retract).

5. When the manual handle of the directional valve is operated, the spool moves to the left position passing the fluid to push and extend the piston against its spring.

5.3 Double-acting hydraulic cylinder

5.3.1 Application

A double-acting hydraulic cylinder shown in Figure 5.2 is used in most of the hydraulic applications where a linear actuation is required in two directions.

5.3.2 Analysis

1. The pump driven by an electric motor extracts the oil from the oil reservoir via a filter or strainer.

2. The pressure relief valve (PRV) allows the oil pressure in the system to reach a specific value after which the PRV opens passing some fluid back to the tank to regulate the pressure.

3. The circuit is controlled by the three-position four-port spring-centered manually-operared directional control valve (DCV).

4. When the DCV is on its center position, there is no fluid flow through it which keeps the piston hydraulically locked in its position.

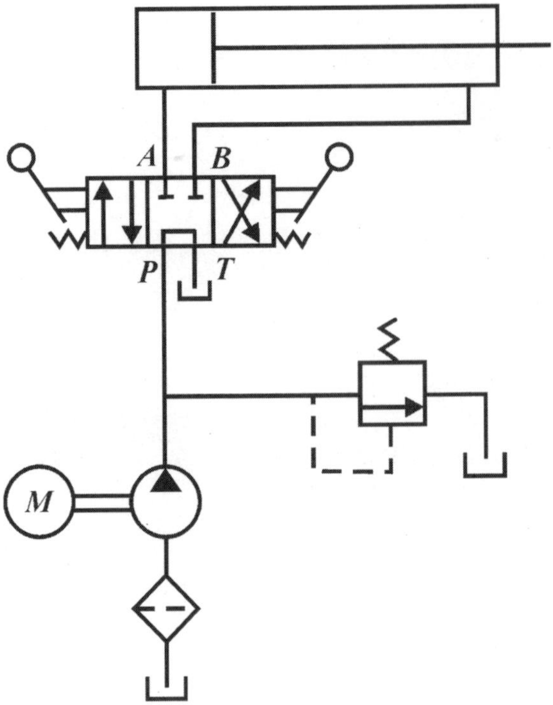

Figure 5.2: *Hydraulic double-acting cylinder*

5. Moving the manual handle of the DCV to the left moves the spool of the valve to the left position, which allows the oil to flow from the pressure line P to port A exerting force on the blank side of the piston extending it to the right. This motion allows the oil to flow from port B coming from the rod side to the tank T.

6. Moving the manual handle of the DCV to the right moves the spool of the valve to the right position, which allows the oil to flow from the pressure line P to port B exerting force on the rod side and retracting the piston to the left. This motion allows the oil to flow from port A coming from the blank side to the tank T.

7. When the piston reaches the dead end on either side, the pressure increases in the pressure line and this can be regulated by the PRV.

5.4 Regenerative hydraulic cylinder

5.4.1 Application

The regenerative circuit shown in Figure 5.3 is usually used to increase the speed of the piston in extension stroke keeping the retraction speed the same as the normal double-acting cylinder. This circuit can be used in shapers to increase the speed of the return stroke (instead of quick return mechanism) and in drilling machines as will be discussed later.

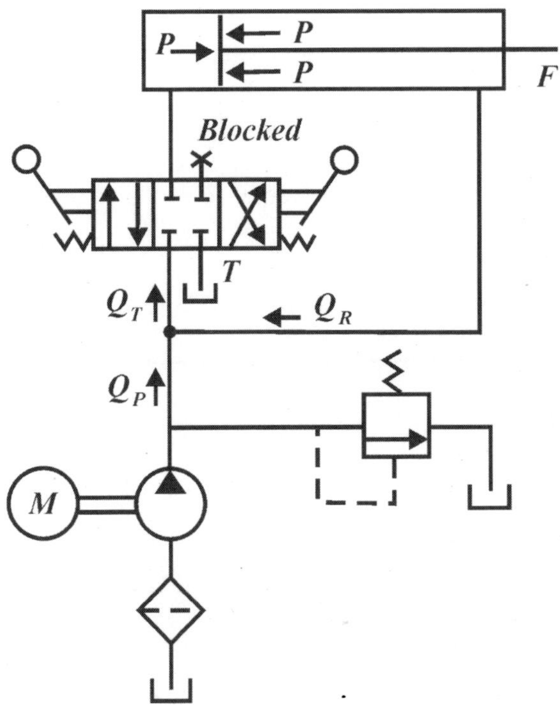

Figure 5.3: *Regenerative hydraulic cylinder circuit*

5.4.2 Analysis

1. The pump extracts oil from the oil tank via a filter or strainer and the PRV aims at keeping specific pressure in the system when the

piston reaches dead ends.

2. One of the output ports of the 3-position 4-way DCV is blocked to allow summing the flow in one port.

3. When the DCV is moved to the right position, the piston retracts normally like a normal double-acting cylinder.

4. When the DCV is set to its left position, the fluid flows to extend the piston to the right sense of direction but with a speed higher than the normal double-acting cylinder because the flow from the rod side Q_R regenerates and sums with the flow of the pump Q_P resulting in a total flow rate Q_T. Increasing the total flow increases the speed of the fluid increasing the piston speed.

Estimating the flow rate of the pump in the extension stroke of the piston:

$$Q_P = Q_T - Q_R$$

Denoting the extension speed of the piston as v, the area of the piston side as A_p and the area of the rod side as $A_p - A_r$, the pump flow rate becomes:

$$Q_P = A_p v - (A_p - A_r)v$$

Solving for the extension speed of the piston gives:

$$v = \frac{Q_P}{A_r}$$

But the force exerted by the regenrative circuit is calculated by:

$$F = PA_r$$

Which is considered lower than the force obtained by normal double-acting cylinder because the rod area is less than the piston area. This leads to the fact that the regenerative circuit increases the speed but decreases the obtained power to carry the load.

One of the main applications on regenerative circuit is the hydraulic drilling machine circuit shown in Figure 5.4.

A hydraulic drilling machine is controlled by a three-position four-way DCV where one of the output ports is regenerated with the other output port. The left position of the DCV causes a normal (slow) extension to operate the drilling feed. The right position causes a normal retraction of the piston after the drilling operation. The center position causes a rapid advance in the extension stroke to approach the drill to the workpiece before beginning the drilling operation to save time in work. A push-button two-position valve is used to begin and stop the operation for better safety condistions.

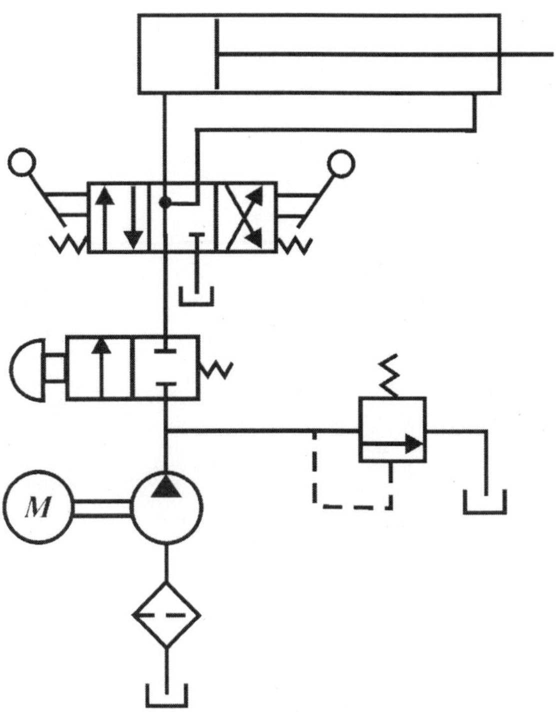

Figure 5.4: *Drilling machine circuit*

5.5 Double-pump hydraulic system

5.5.1 Application

The circuit shown in Figure 5.5 is based on using a high-pressure low-flow pump and another low-pressure high-flow pump to feed the system with oil. This circuit is used in general to design a hydraulic punch press for shearing and forming of sheet metals. The main purpose of this circuit is to introduce a rapid extension stroke of the piston under low pressure for the free of load stroke and a slow retraction stroke with high pressure to exert high force when the punch approaches the workpiece to execute the cutting or forming operation.

5.5.2 Analysis

1. The high-flow low-pressure pump delivers oil to the circuit when there is no punching load on the piston in the extension stroke.

2. The three-position four-port DCV is used to control the direction of operation.

3. When the punch approaches the workpiece, the high punching load increases the pressure in the system which opens the unloading PRV by the pilot line eliminating the influence of the high-flow pump and operating the low-flow high-pressure pump only to exert higher force on the load executing the punching operation.

4. As soon as the punching operation is finished, the pressure in the system reduces leading the pilot line to close the unloading PRV and to return the effect of the high-flow pump and increasing the speed of the punch in the retraction stroke.

5.6 Locked cylinder hydraulic system

5.6.1 Application

In most of the applications where the hydraulic system is in contact with with human beings, there is a need to lock the cylinder in its position

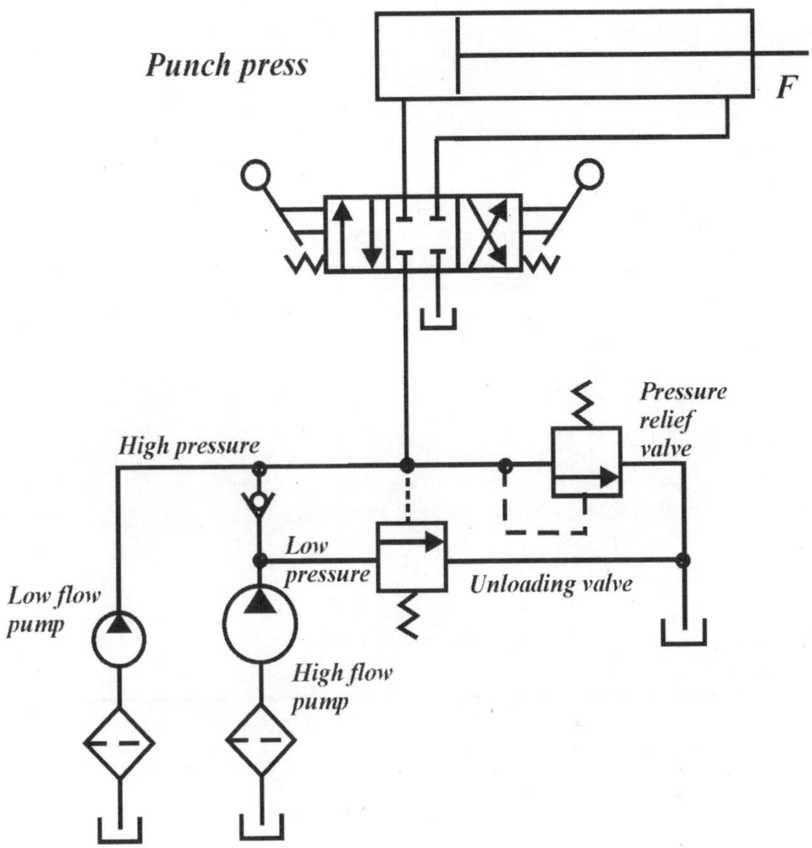

Figure 5.5: *Double pump hydraulic system (Punch press)*

to avoid causing harm to people. This can be done by using pilot check valves as shown in Figure 5.6. Examples of these applications are lifts, cranes, concrete pumps, fork lifts and loaders.

5.6.2 Analysis

1. Oil is pressurized into the system by a pump fed by an oil reservoir via an oil filter. A PRV is used here to regulate and avoid the overload pressure in the circuit.
2. Setting the three-position four-way DCV to the left position causes an increase in the pressure line to extend the piston.

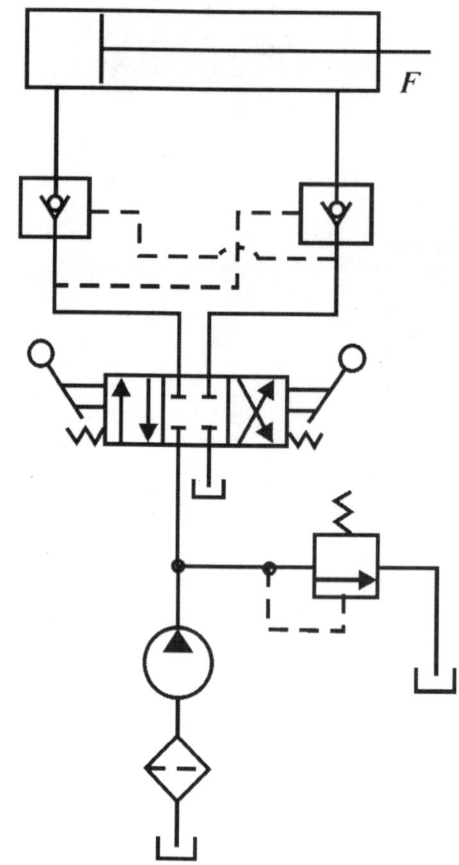

Figure 5.6: *Locked cylinder hydraulic system using check valves*

3. Oil can pass freely from the left hand check valve but the right hand check valve resists this flow.

4. When pressure increases in the left line, the pilot line opens the right check valve permitting the fluid to flow and allowing the piston to extend.

5. changing the DCV to the right position, reverses the operation causing the left check valve to open and the piston to retract.

6. when the DCV is on its middle spring-centered position, the piston is locked in its place without being able to extend or retract under any external load.

5.7 Counterbalance hydraulic system

5.7.1 Application

A counterbalance installation (shown in Figure 5.7) is used to protect a vertically mounted hydraulic cylinder from moving under vertical loads when the pump is idling. This system is used when a load is hanged in an upward position to a hydraulic cylinder.

5.7.2 Analysis

1. Oil is pressurized into the system by the pump fed by an oil reservoir via an oil filter. A PRV is used here to regulate and avoid the overload pressure in the circuit.

2. The system is controlled by a three-position four-way spring-centered solenoid-controlled DCV.

3. Moving the spool to the left position allows the oil to flow through the check valve to the rod side of the piston causing retraction operation upwards. The required pressure here is to overcome the load.

4. Changing the DCV to the right position passes the flow to the blank side of the piston to extend it. The oil flow out of the piston is

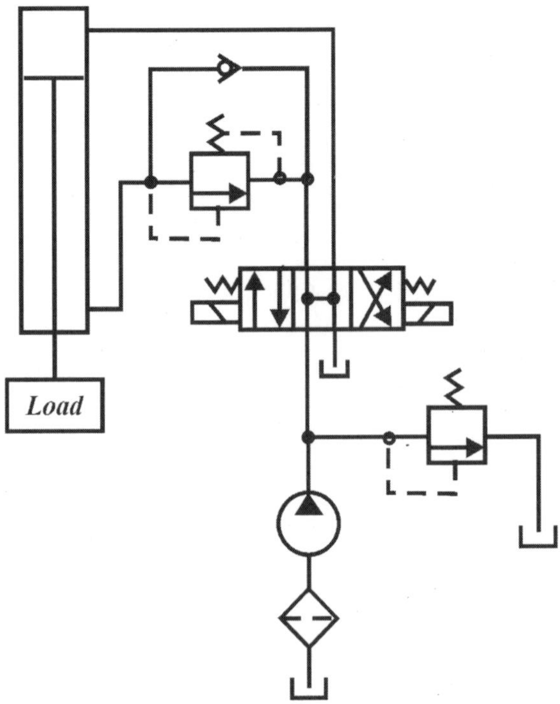

Figure 5.7: *Counterbalance hydraulic system*

restricted by the pilot PRV. A specific value of pressure is needed to do this operation. Thus, the PRV is set to a pressure value that can overcome the load with a small overhead.

5. When the DCV is on center position, all lines are open to the tank including the pressure line coming from the pump which causes the pump to circulate oil to the tank. In this position, the load is kept hanged by the pressure of the PRV that behaves like a counterbalance.

5.8 Sequence cylinder hydraulic system

5.8.1 Application

The cylinder sequencing circuit shown in Figure 5.8 is used for executing operations in sequence like folding the sides of a metal scrap press or packaging and arranging boxes in warehouses.

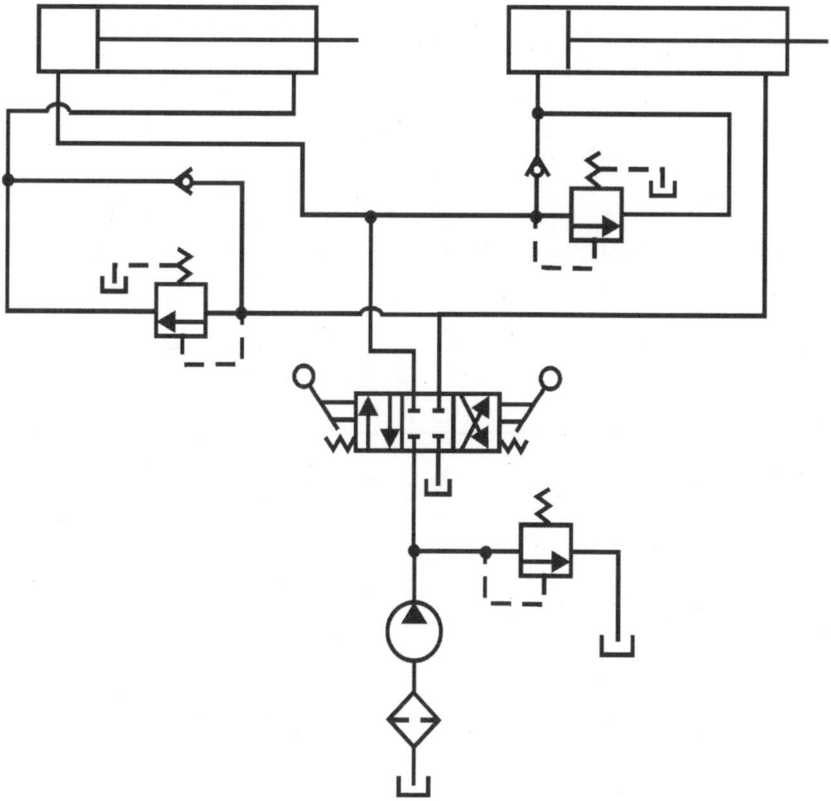

Figure 5.8: *Sequence cylinder hydraulic system*

5.8.2 Analysis

1. Oil is pressurized into the system by the pump fed by an oil reservoir via an oil filter. A PRV is used here to regulate and avoid the overload

pressure in the circuit.

2. When the DCV is moved to the left position, the oil selects to pass to the left cylinder which is the easiest way leading to full extension of this cylinder.

3. Reaching the dead end (full stroke) of the left cylinder increases the oil pressure in this line which opens the PRV at the inlet of the right cylinder and extends the right cylinder.

4. From the previous two steps 2 and 3, one can see that the operation is executed by extending the left piston first and then the second piston in sequence.

5. The retraction operation occurs when the spool of the DCV is pushed to the right position which causes retracting the right cylinder first followed by the retraction of the left cylinder in sequence.

6. In each operation, the oil returns back to the tank through the check valve installed on the by-pass line passing around the PRV.

7. The spring centered mid position of the DCV blocks the two cylinders in their positions.

5.9 Automatic reciprocating hydraulic system

5.9.1 Application

The automatic reciprocating system shown in Figure 5.9 can be used to function a double-acting hydraulic cylinder back and forth automatically. This system is used to operate the pumping reverser in some concrete pumps and in the automatic reciprocating motion of the table of a surface finishing machine.

5.9.2 Analysis

1. Oil is pressurized into the system by the pump fed by an oil reservoir via an oil filter. A PRV is used here to regulate and avoid the overload pressure in the circuit.

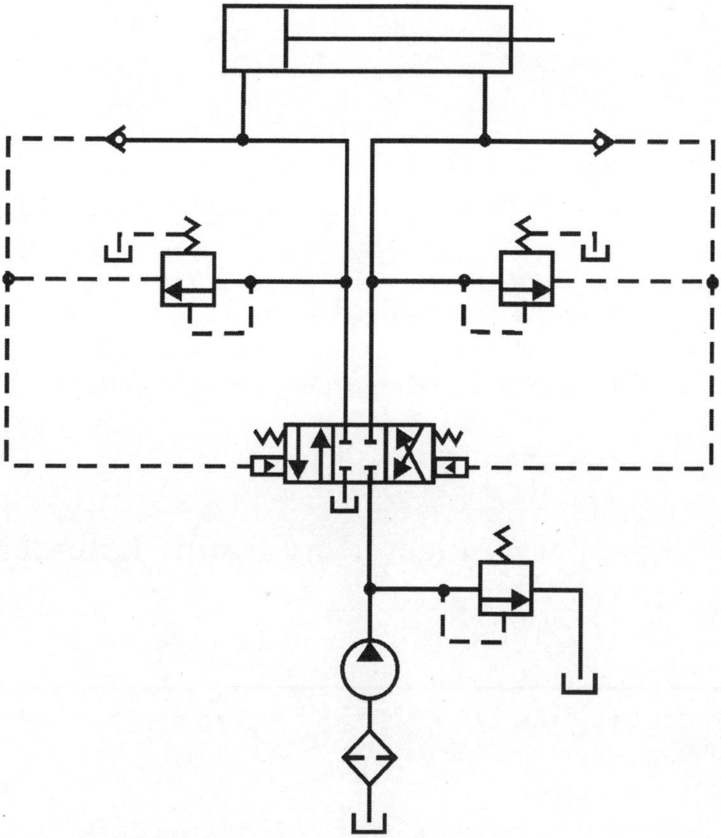

Figure 5.9: *Automatic reciprocating hydraulic system*

2. A three-position, four-way spring-centered and pilot-operated DCV is used to control this circuit.

3. Leaving the DCV on the center position keeps the cylinder blocked in its position.

4. The reciprocating operation begins by moving the DCV to any of the right or left positions.

5. Beginning from the right position of the DCV; the pressurized oil flows in the left line and extends the cylinder till reaching full stroke.

6. At full extension stroke, the pressure builds up in the left line and opens the left PRV sending oil via the pilot line to the left pilot control in the DCV moving the spool to the left position.

7. The spool now is on the left position of the DCV; the pressurized oil flows in the right line and retracts the cylinder till reaching zero stroke.

8. At complete retraction, the pressure builds up in the right line and opens the right PRV sending oil via the pilot line to the right pilot control in the DCV moving the spool to the right position.

9. And so on and so forth ...

5.10 Parallel connected hydraulic cylinders

5.10.1 Application

Having an identical motion of two cylinders in synchronization can be obtained by connecting the two cylinders in parallel as shown in Figure 5.10. Two conditions must be satisfied to obtain this synchronization:

- The two cylinders have to be completely identical.
- The two loads acting on the cylinders have to be equal.

It is worth mentioning that it is impossible to have two identical cylinders because of the differences in friction and accuracy in manufacturing. On the other hand, it is not easy to get two equal loads on the cylinders. Nevertheless, cylinders in parallel are used specially in loader bucket and heavy booms that need two pistons to act at the same point.

5.10.2 Analysis

1. The pump, tank, filter and PRV have the same function as in previous circuits.

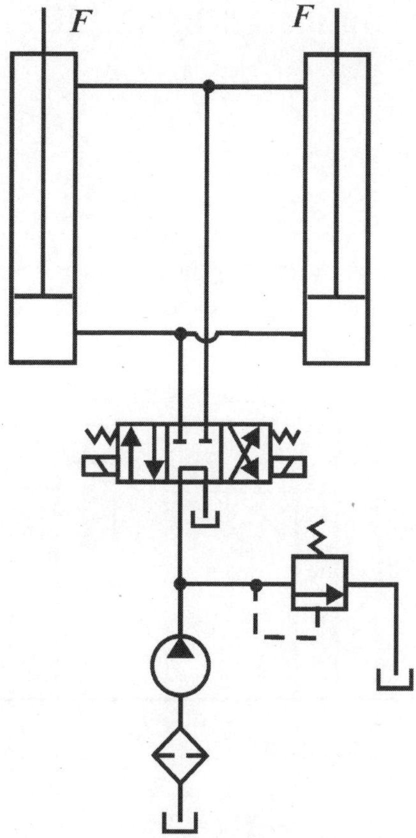

Figure 5.10: *Hydraulic cylinders connected in parallel*

2. leaving the three-position four-way solenoid-controlled DCV on the center position by the influence of the springs causes the pump to circulate the oil to the tank without influencing the system.

3. Moving the DCV to the left position prssurizes the oil to extend the two pistons simultaneously if they are identical and if the two loads are equal.

4. Moving the DCV to the right position pumps the oil to retract the two pistons under the same conditions.

5.11 Series connected hydraulic cylinders

5.11.1 Application

The best way to obtain complete synchronization for two cylinders regardless of the acting loads is connecting them in seris as shown in Figure 5.11. The condition for this connection is to have the blank area of the second cylinder equals to the difference between the blank area and the rod area of the first cylinder.

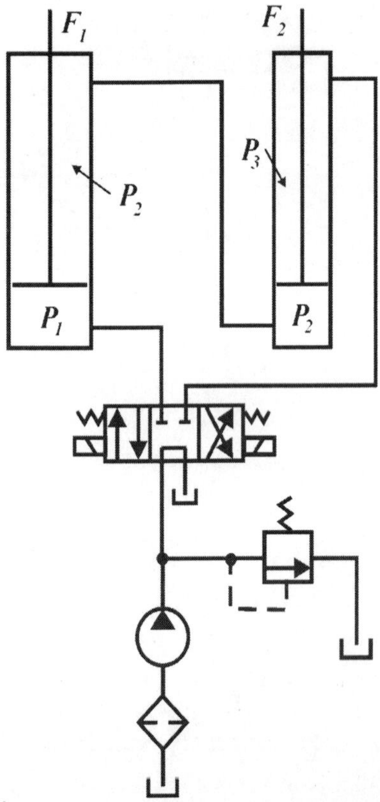

Figure 5.11: *Hydraulic cylinders connected in series*

5.11.2 Analysis

1. The pump, tank, filter and PRV have the same function as in previous circuits.

2. leaving the three-position four-way solenoid-controlled DCV on the center position by the influence of the springs causes the pump to circulate the oil to the tank without influencing the system.

3. Setting the DCV to the left position pushes the fluid to the blank end of piston 1 to extend this piston with a pressure P_1. The oil in the rod side of piston 1 is drained with a pressure P_2 to become the inlet to piston 2 with the same pressure. Finally, the fluid in the rod side of piston 2 is drained back to the tank at a new pressure P_3.

Applying the continuity equation and denoting Q for the flow rate, A for the area and v for the velocity with the subscripts 1 and 2 for cylinders 1 and 2 respectively.

$$(Q_{out})_1 = (Q_{in})_2$$

Since $Q = Av$,

$$(Av)_1 = (Av)_2$$

or,

$$((A_P)_1 - (A_R)_1)v_1 = (A_P)_2 v_2$$

Assuming that the fluid is incompressible, the synchronization means that $v_1 = v_2$, then

$$(A_P)_1 - (A_R)_1 = (A_P)_2$$

Note that the pressure P_1 is the total pressure needed to carry the two loads and the output pressure P_3 is zero gauge unless there is a PRV at the return line to the tank. Newtons second law can be applied for the summation of the force on cylinder 1:

$$P_1(A_P)_1 - P_2((A_P)_1 - (A_R)_1) = F_1 \tag{5.1}$$

Summing the forces on cylinder 2 results

$$P_2(A_P)_2 - P_3((A_P)_2 - (A_R)_2) = F_2 \tag{5.2}$$

But $P_3 = 0$ and $(A_P)_2 = (A_P)_1 - (A_R)_1$, then

$$P_2((A_P)_1 - (A_R)_1) = F_2 \tag{5.3}$$

Summing equations (5.1) and (5.3) gives

$$P_1(A_P)_1 = F_1 + F_2 \tag{5.4}$$

5.12 Fail-safe circuit

5.12.1 Application

The fail-safe circuit drawn in Figure 5.12 is used to protect the machine from accidentally falling on the operator in machines working in contact directly with people. It is used to avoid any damage or overload from hurting operators.

5.12.2 Analysis

1. The pump, filter, tank and PRV are used as usual for all the previous circuits.
2. A two-position four-way, pilot-controlled DCV is used here.
3. When the DCV is set to the righ position by means of the spring, the piston retracts to its end position.
4. The piston is prevented from extending by the effect of the check valve installed on the return line unless the push-button valve is pressed.
5. when the push-button valve is pushed, it operates the DCV to extend the piston after opening the check valve through the pilot line.

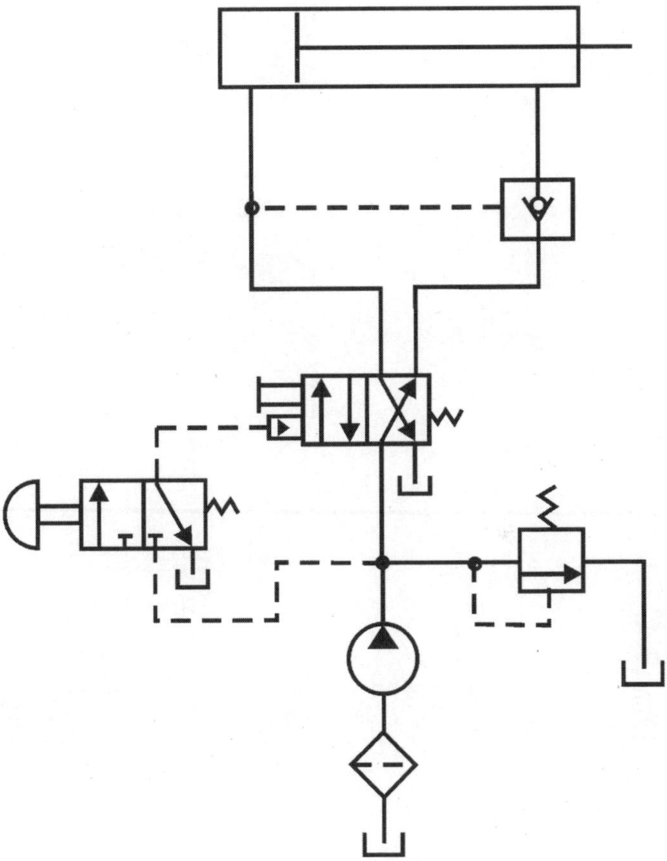

Figure 5.12: *Hydraulic fail-safe circuit*

Figure 5.13 shows a two-handed safety circuit. For further protection, the operator needs to push the two push-buttons to pilot operate the DCV to extend the piston. Otherwise, the piston will stay retracted no matter if the operator pushes one button or nothing.

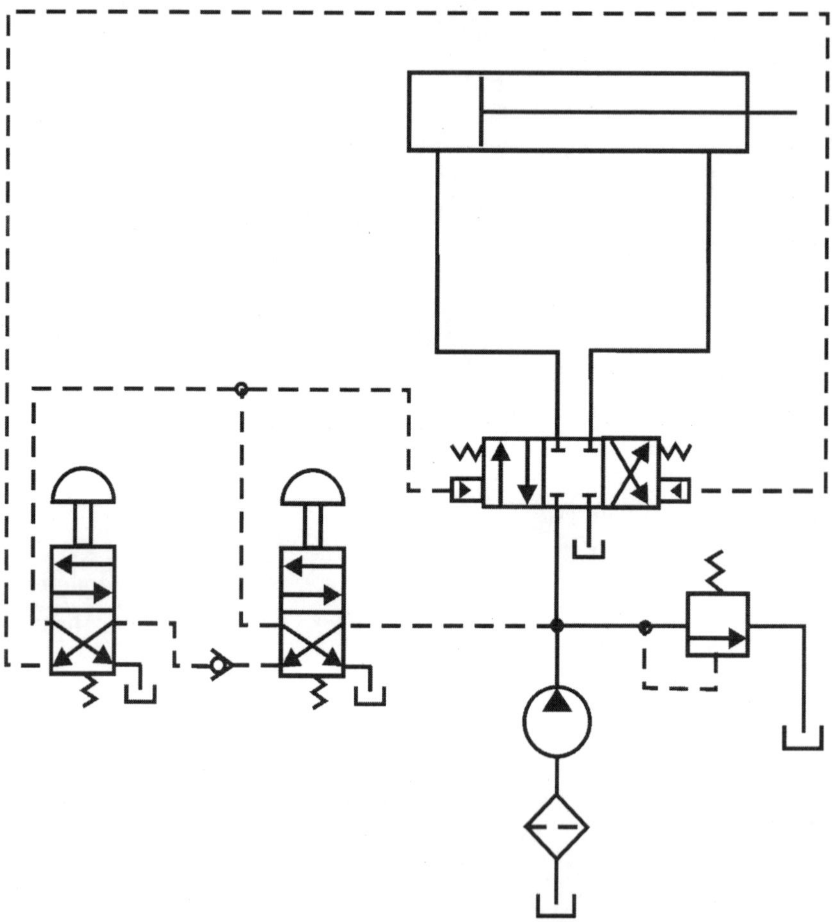

Figure 5.13: *Two-handed safety circuit*

5.13 Meter-in cylinder speed control

5.13.1 Application

Meter-in speed control (Figure 5.14) is used to regulate the speed of the hydraulic pistons by controlling the flow rate of the oil entering to the cylinder when the speed is significant in specific applications.

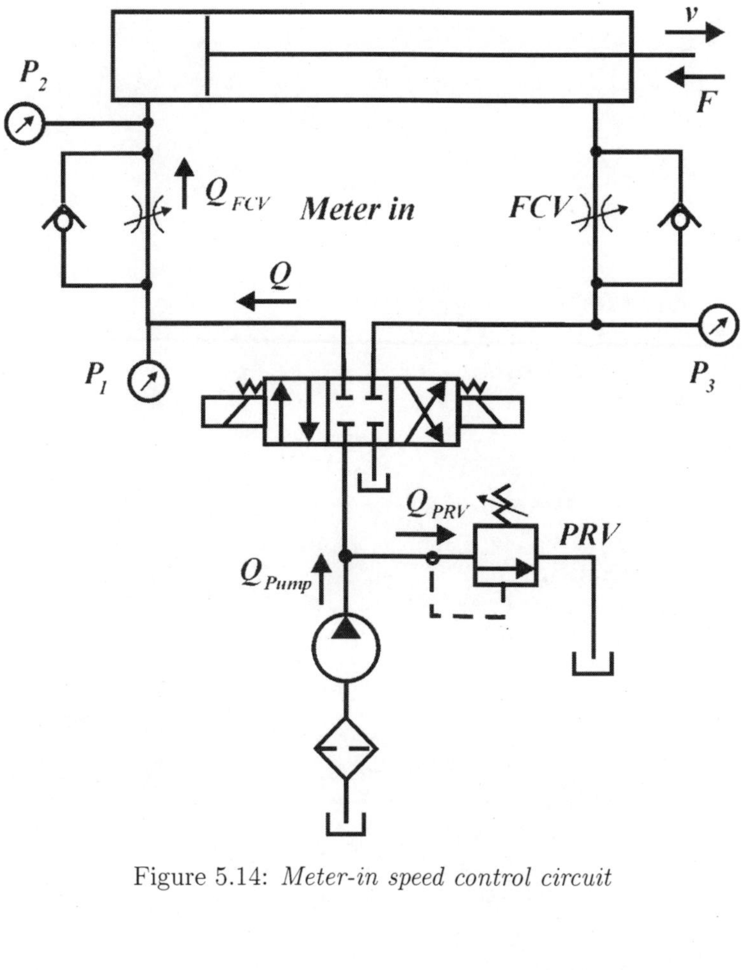

Figure 5.14: *Meter-in speed control circuit*

5.13.2 Analysis

1. The pump, filter, tank and PRV are used as usual for all the previous circuits.

2. This circuit works exactly like the operation of a double-acting hydraulic cylinder with the difference of installing flow control valves (FCV) on the hydraulic lines by-passed by check valves.

3. The inlet oil is forced to pass through the FCV to control the speed while at the outlet, it can pass through the check valve without restrictions.

4. Controlling the flow rate at the piston inlet controls the speed of the piston.

5. One disadvantage of the meter-in design is that it cannot control an overrunning load.

5.14 Meter-out cylinder speed control

5.14.1 Application

Meter-out speed control (Figure 5.15) is also used to regulate the speed of the hydraulic pistons by controlling the flow rate of the oil exiting the cylinder when the speed in significant in specific applications.

5.14.2 Analysis

1. The pump, filter, tank and PRV are used as usual for all the previous circuits.

2. This circuit also works exactly like the double-acting hydraulic cylinder with the difference of installing flow control valves (FCV) on the hydraulic lines by-passed by check valves too.

3. The outlet oil is forced to pass through the FCV to control the speed while at the inlet it can pass through the check valve without restrictions.

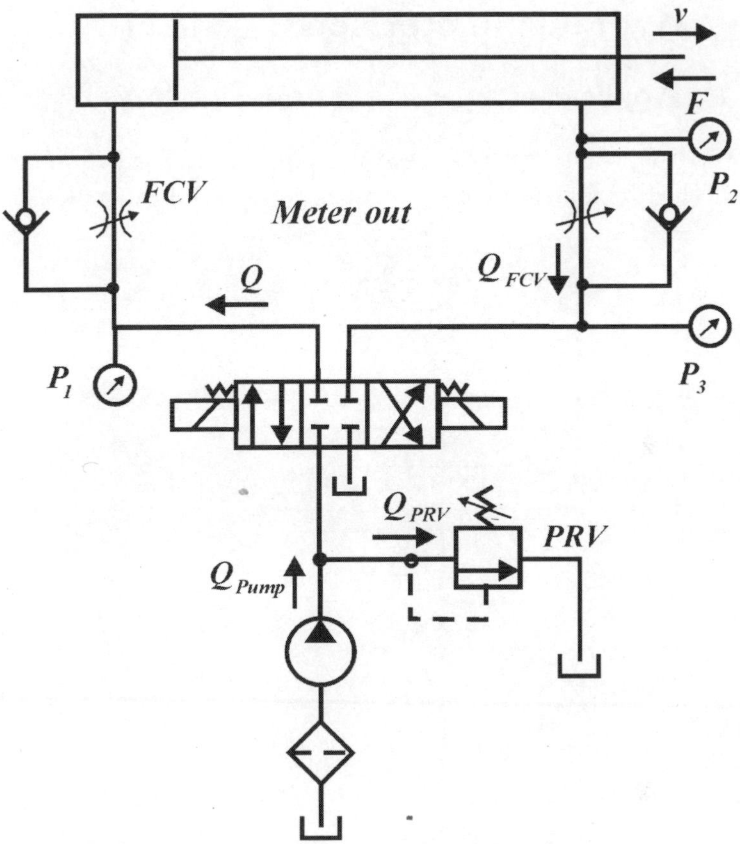

Figure 5.15: *Meter-out speed control circuit*

4. Controlling the flow rate at the piston outlet controls the speed of the piston.

5. The disadvantage of meter-out control is that the pressure builds up excessively in the rod side of the piston which increases the temperature reducing the efficiency. Despite that meter-out is more efficient than meter-in design.

5.15 Hydraulic motor speed control

5.15.1 Application

Figure 5.16 depicts an open-loop speed control circuit of a hydraulic motor accomplished by a pressure-compensated flow control valve FCV.

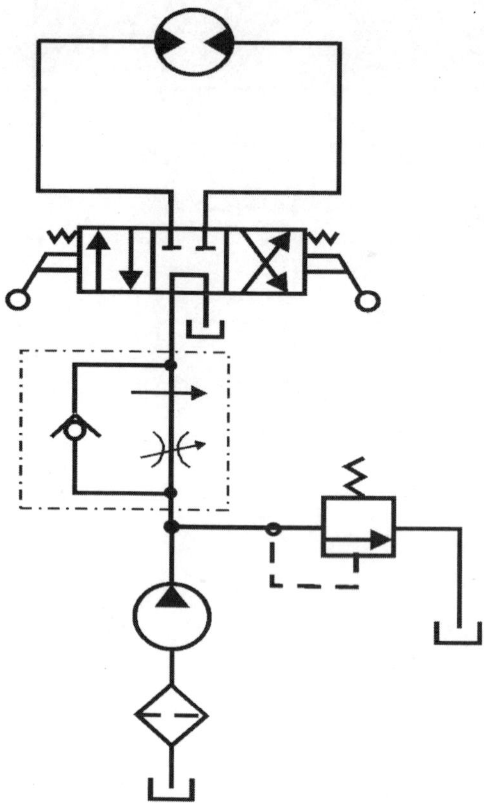

Figure 5.16: *Pressure compensated speed control for a hydraulic motor*

5.15.2 Analysis

1. The pump, filter, tank and PRV are used as explained for all the previous circuits.

2. When the three-position four-way DCV is spring-centered, the pump circulates the oil in an idle manner and the hydraulic motor is hydraulically blocked.

3. Pushing the DCV to the left position passes the oil to rotate the hydraulic motor in one direction.

4. The speed of the motor is controlled by changing the throttle of the FCV and if the there is an excess pressure according to the load it can be drained by the PRV.

5. Pushing the DCV to the right position passes the oil to rotate the hydraulic motor in the other direction.

6. The by-pass check valve aims at compensating and equalizing the pressure on the two sides of the FCV.

5.16 Hydraulic motor braking system

5.16.1 Application

The circuit in Figure 5.17 shows an open-loop installation of a two-directional hydraulic motor with braking system. This is used when the motor operates a high inertia load applications.

5.16.2 Analysis

1. The oil is pressurized after being drawn from the tank by the pump and the DCV changes the direction of rotation of the hydraulic motor by changing position manually.

2. When the DCV is spring centered to mid position the hydraulic motor is blocked. If the motor is driving a high inertia load, this can cause flywheel effect on it which forces it to move slightly pushing fluid to one side. This fluid will be drained back to the tank via the PRV installed between the two check valves on the right.

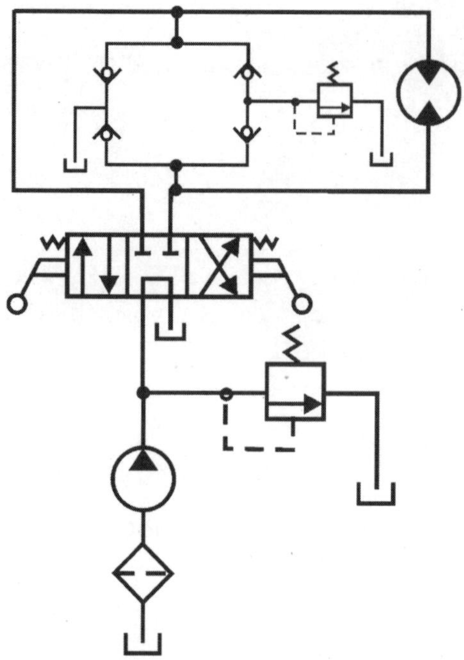

Figure 5.17: *Braking system for a hydraulic motor*

3. Leaking fluid caused by high inertia reduces the amount of fluid in the circuit. To avoid pulling air, the other line with two check vàlves on the left, pumps oil from the tank into the circuit.

5.17 Hydrostatic transmission system

5.17.1 Application

The closed-loop circuit of a reversible variable displacement pump and two-directional hydraulic motor is shown in Figure 5.18. This is usually called "Hydrostatic Transmission System". One of the most common applications for this circuit is the concrete mixer.

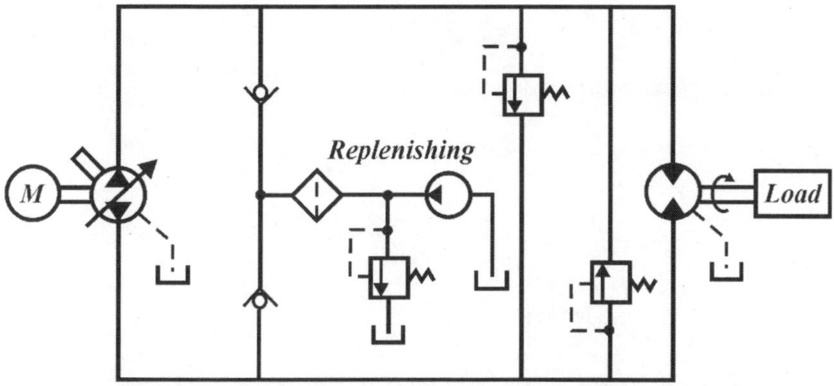

Figure 5.18: *Hydrostatic transmission system*

5.17.2 Analysis

1. The system is driven by a reversible variable displacement hydraulic pump driven by an electric motor.

2. Changing the direction of the pump changes the direction of rotation of the hydraulic motor.

3. The motor speed is controlled infinitly by varying the flow rate (displacement) of the pump.

4. When the motor is overloaded, the oil is circulated through one of the overload PRVs depending on the direction of rotation without damaging the system.

5. When the motor is influenced by the high inertia of the load, part of the fluid is leaked to the tank and is replenished from the tank to the system by one of the replenishing check valves according to the direction of rotation.

5.18 Air over oil system

5.18.1 Application

Figure 5.19 depicts an air over oil circuit where the air is used instead of the hydraulic pump. This kind of circuits is used in vehicle lifting jacks.

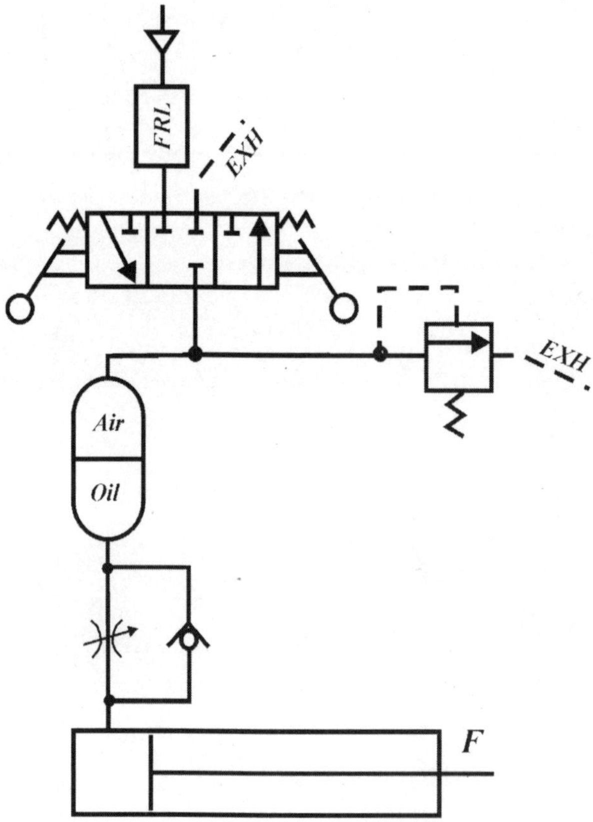

Figure 5.19: *Air over oil system*

5.18.2 Analysis

1. The compressed air flows coming from an air storage (surge) tank fed by a compressor.

2. The air passes through a Filter Regulator Lubricator (FRL) to a three-position, three-way DCV.
3. The circuit is supplied with a PRV to exhaust the air out of the circuit to the atmosphere in the case of builtup pressure.
4. The compressed air enters into a vessel in which it exerts pressure on an oil separated by a diaphram.
5. The oil, in turn, passes to extend the hydraulic pistion to lift the load.
6. When the piston retracts under the load, the oil passes through a FCV to control its return speed to avoid impact.

5.19 References

[1] A. Esposito, *Fluid Power with Applications*, Prentice Hall, 2003.

[2] A. Preumont, *Vibration control of active structures*, 1st. ed., Press Universitaire de Bruxelles, pp33-34, 1996-1997.

[3] M. Pinches and J. Ashby, *Power Hydraulics*, Prentice Hall, 1989.

Chapter 6

Pneumatic Circuits

6.1 Introduction

Pneumatic solution is considered the most common relatively high power and low cost solution to supply systems with power and motion. The circuits discussed in this chapter concentrate on the systems controlled by mechanical and electro-mechanical valves and controls. A general pneumatic circuit consists mainly of:

- Air compressor (piston, screw or rotary type).
- Air reservoir (tank to be filled with compressed air).
- Filter-Regulator-Lubricator (FRL).
- Pneumatic actuator (linear piston or rotary motor).
- Pneumatic valve (pressure relief, flow control or directional control).

The same considerations taken in hydraulic circuits must be taken into account when designing a pneumatic system, these considerations are:

- Saftey of operator and operation.
- Efficiency and performance of the whole system.
- Cost requirements.
- Simplicity and easiness.

Different pneumatic circuits will be discussed in the following sections using the fluid power symbols listed before and that conform with the Amercian National Standards Institute (ANSI).

6.2 Single-acting pneumatic cylinder

6.2.1 Application

The single-acting cylinder shown in Figure 6.1 is used when there is a need to have an action in a single sense of direction while the return stroke of the cylinder is actuated by means of a spring. An example of application is the pushers and rejectors used to reject manufactured plastic vessels that fail the leakage test in a production line.

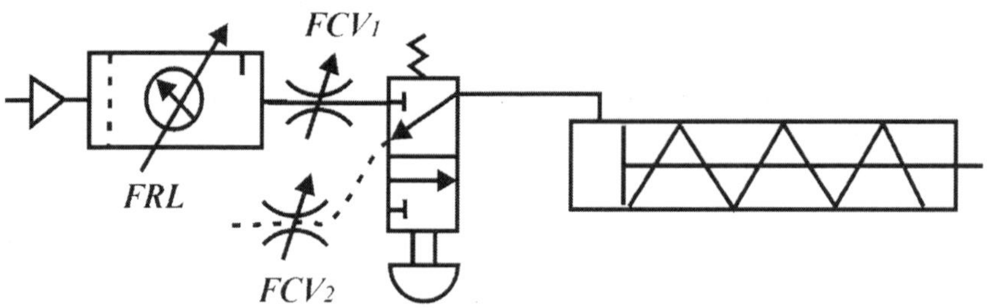

Figure 6.1: *Pneumatic single-acting cylinder*

6.2.2 Analysis

1. The inlet to the system begins from compressed air coming from the air storage tank after being pressurized by a compressor.

2. The compressed air passes through a Filter Regulator Lubricator (FRL) in which it is filtered and lubricated by adding some oil drops. The FRL serves also to regulate the air pressure before being fed to the application.

3. If the button of push-button directional valve is pressed the air flow is allowed to pass through the valve to extend the piston. The speed

of the extension stroke of the piston is controlled by the Flow Control Valve (FCV1).

4. Releasing the push-button valve returns the valve to the spring loaded position and opens the port of the piston to the exhaust port.

5. The psiton is returned back by the spring force and the air is exhaused through the Flow Control Valve (FCV2) that controls the speed of the retraction stroke.

6.3 Double-acting pneumatic cylinder

6.3.1 Application

Figure 6.2 depicts a douple-acting piston which is used in general for most of the applications that need actions in the two senses of direction; extension and retraction.

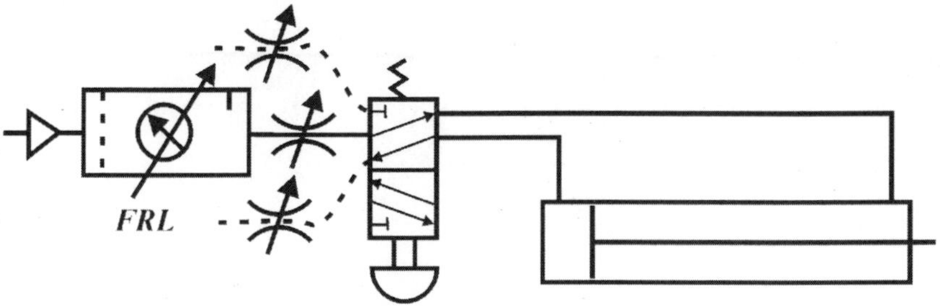

Figure 6.2: *Pneumatic double-acting cylinder*

6.3.2 Analysis

1. The inlet to the system begins from compressed air coming from the air storage tank after being pressurized by a compressor and passed through an FRL.

2. A two-position five-port push-button directional control valve is used to change the direction of motion of the piston.

3. Pushing the button of the directional valve passes the compressed air to extend the piston and the excess air is vented through the exhause port of the valve.

4. Releasing the button returns the control valve to the spring loaded position that allows the compressed air to retract the piston back exhausting the air from the exhaust port of the valve.

5. The speeds of the extension and retraction strokes are controlled by the variable flow control valves installed on the different ports.

6.4 Air pilot control of pneumatic cylinder

6.4.1 Application

The air pilot control can be used to change the direction of actuation of the pneumatic piston as shown in Figure 6.3. This circuit is used to limit the stroke of the piston at both sides like the one used in the reciprocating table of the grinding machine.

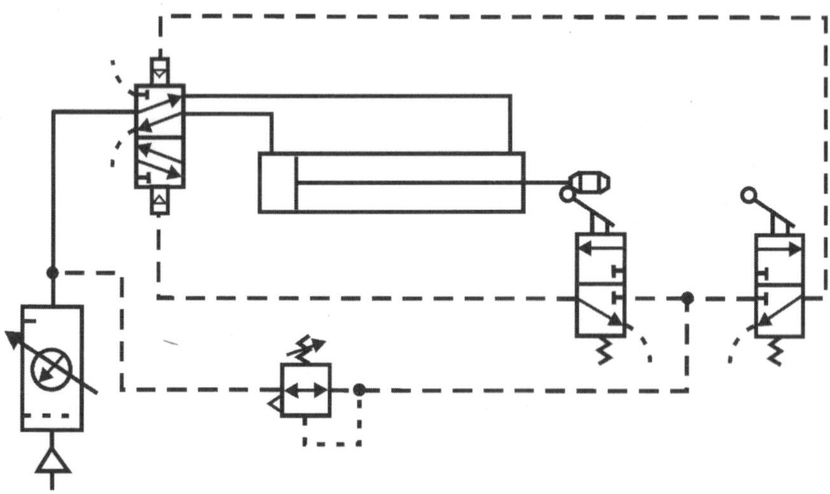

Figure 6.3: *Air pilot control of a pneumatic double-acting cylinder*

6.4.2 Analysis

1. The inlet to the system begins from compressed air coming from the air storage tank after being pressurized by a compressor and passed through an FRL.

2. A two-position five-way pilot-actuated DCV is used to change the direction of actuation of the piston by passing the high pressure air to the piston everytime when the position is changed.

3. An air pressure relief valve is used to reduce the pressure (sometimes to 10% of the actuation pressure) to be used for the pilot operation that controls the position of the DCV.

4. When the piston reaches the end of the extension stroke, the tip of the piston pushes the handle of the push-button valve changing its position and passing the pilot air to the second position of the DCV. This changes the position of the DCV and passes the pressurized air to retract the piston.

5. When the piston retracts to the end of the stroke it pushes the handle of the second push-button valve repeating the same operation and extending the piston again, and so on and so forth.

6. This circuit leads to a reciprocating motion of the cylinder.

6.5 Cycle timing of pneumatic cylinder

6.5.1 Application

The circuit shown in Figure 6.4 is used for the timing of the cylinder cycle where a limited stroke is needed after which the piston is retracted again.

6.5.2 Analysis

1. The inlet to the system begins from compressed air coming from the air storage tank after being pressurized by a compressor and passed through an FRL.

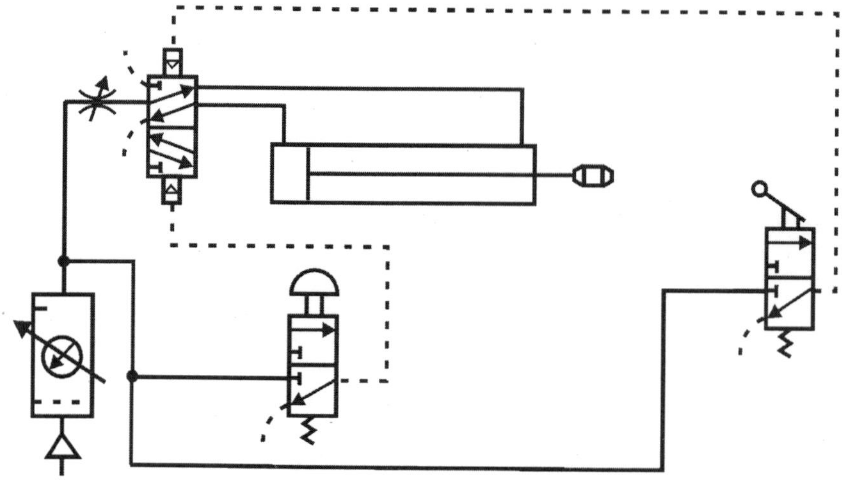

Figure 6.4: *Cycle timing of pneumatic cylinder*

2. When the push-button valve is pressed manually, the position of the two-position five-way valve is set to extend the stroke of the piston.

3. The tip of the piston presses the handle of the limit valve to return the piston back in a retraction stroke.

4. To make another cycle of extension and retraction, the push-button valve must be pressed manually again.

5. The FCV is used to control the extension and retraction speed of the piston.

6.6 Two-speed pneumatic cylinder

6.6.1 Application

Figure 6.5 shows a circuit serves to introduce two different speeds for a piston during the same stroke. This circuit is used in cutting and forming press when there is a need to accelerate the piston before reaching the load and deccelerating it during the forming operation to increase the force.

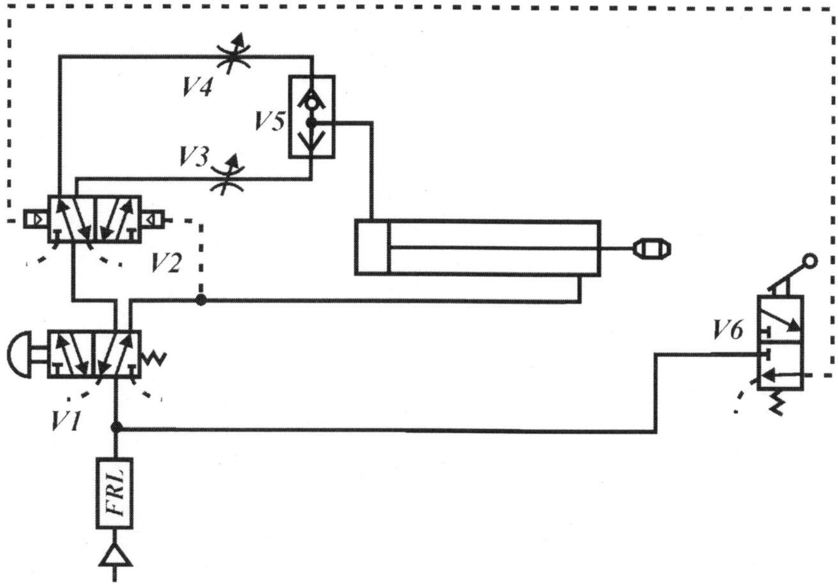

Figure 6.5: *Two-speed pneumatic cylinder*

6.6.2 Analysis

1. The system is fed by a compressed air via an FRL.

2. Pressing the button of the push-button valve V1 passes the compressed air through the DCV V2 and the FCV V4 to the shuttle valve V5 to begin extending the piston in a high speed depending on the flow controlled by the FCV V4.

3. When the tip of the piston presses the handle of valve V6, the pilot air flows to change the position of the DCV V2. This operation passes the compressed air through the FCV V3 via the shuttle valve to continue extending the piston but with a different lower speed depending on the flow controlled by the FCV V3.

4. Releasing the push-button valve V1 retracts the piston to the original position.

6.7 Two-handed safety circuit for pneumatic cylinder

6.7.1 Application

Figure 6.6 shows a safety circuit used to hold the piston in the retraction position unless the two push-button valves are pressed. This circuit is necessary in the applications close to human beings to avoid injuries caused by unconcious operating of the cylinder using one single button.

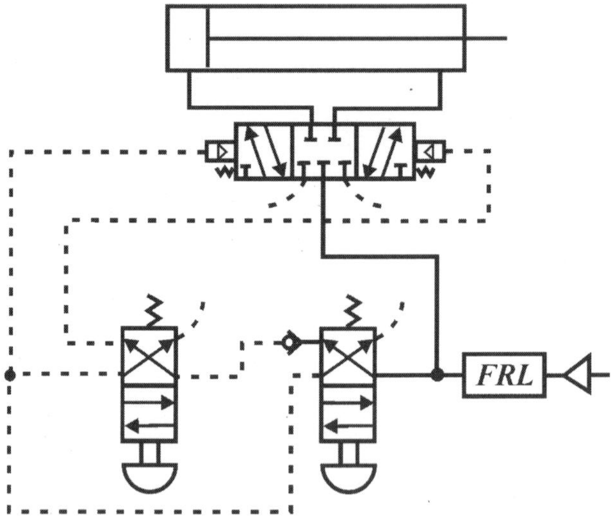

Figure 6.6: *Two-handed safety circuit for pneumatic cylinder*

6.7.2 Analysis

1. The system is fed by a compressed air via an FRL.

2. Pushing the two palm-button valves passes the compressed air to right side envelope of the three-position five-way pilot-actuated DCV causing the piston to extend.

3. Note that pushing one valve will exhaust the air out and will not actuate the piston.

4. Releasing any of the pushed valves will return the DCV to the left position causing the piston to retract.

5. safety here comes from the necessity to push both palm-button valves to extend the piston.

6.8 Control of air motor

6.8.1 Application

The circuit shown in Figure 6.7 represents an operation procedure for an air motor where one push-button is used to actuate the motor and the other push-button is used to stop it.

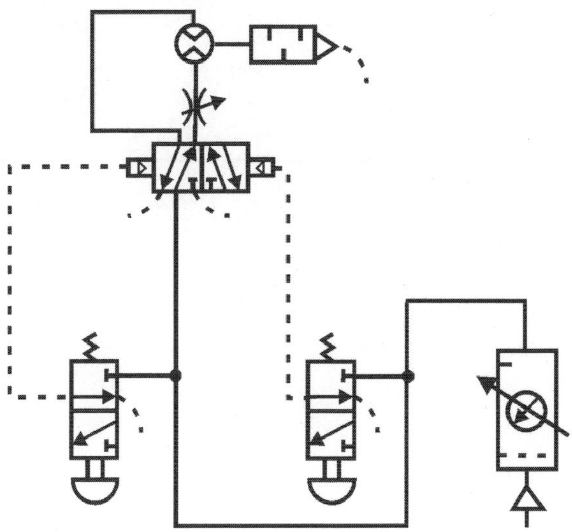

Figure 6.7: *Control of air motor*

6.8.2 Analysis

1. The system is fed by a compressed air via an FRL.

2. Pushing the left hand push-button will pass the compressed air to actuate the air motor by changing the position of the two-position five-way pilot-operated DCV.

3. The rotational speed of the air motor is controlled by variable flow control valve.

4. Pushing the right hand push-button will change the position of the DCV leading to vent the air out and to stop the air motor.

6.9 Deceleration of a pneumatic cylinder

6.9.1 Application

Figure 6.8 shows a circuit used to decelerate the speed of the air piston when it is exposed to high loads to avoid shocks at the ends of the strokes.

6.9.2 Analysis

1. The system is fed by a compressed air via an FRL.

2. When there is a large weight acting on the piston in the retraction stroke, the pilot air passes through FCV V3 and changes the position of DCV V5 to force the exhausting air to be restricted by FCV V7 and decelerating the cylinder.

3. The same process is repeated in the extension stroke through valves V6, V2 and V4.

6.10 Gas loaded accumulator

6.10.1 Application

Figure 6.9 shows how to use a gas loaded accumulator to give high impact to the motion of the piston. An example of this circuit is used in the pumping system of a concrete pump.

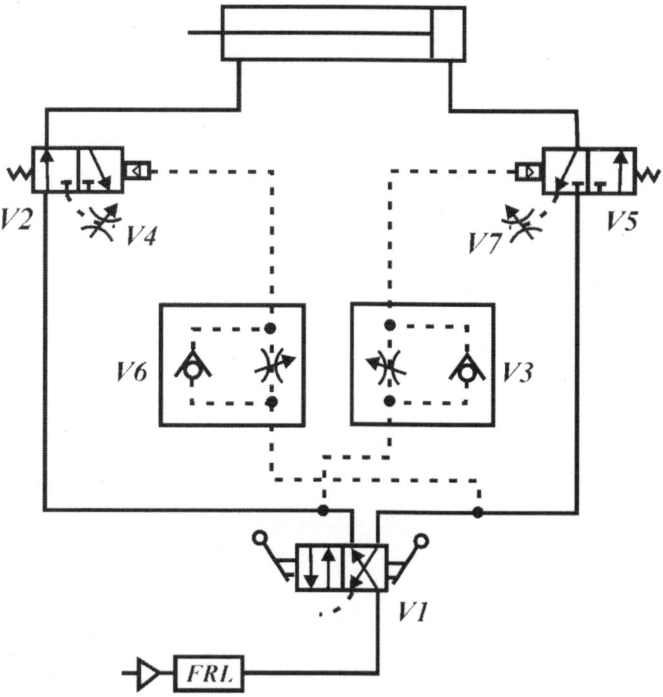

Figure 6.8: *Deceleration of a pneumatic cylinder*

6.10.2 Analysis

When the piston reaches a dead end, the accumulator is filled with oil compressed by the loaded gas. Changing the direction of the stroke begins the action with a high impact caused by the accumulator to overcome any blockage or high load effect.

6.11 References

[1] A. Esposito, *Fluid Power with Applications*, Prentice Hall, 2003.

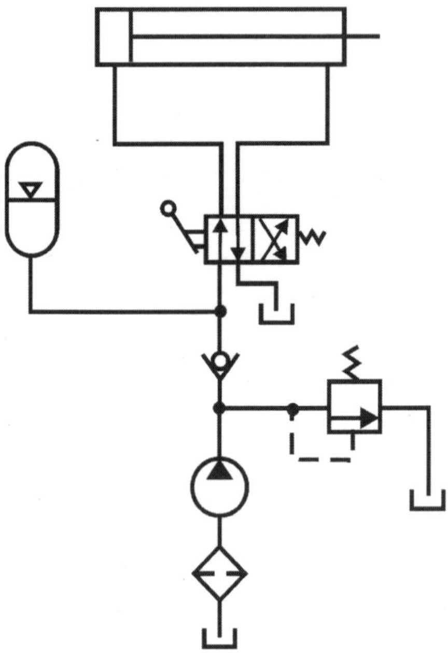

Figure 6.9: *Gas loaded accumulator*

Chapter 7

Fluid Power Components

7.1 Introduction

Any fluid power system consists mainly of the following components:

- Prime mover (Electrical or Internal Combustion motor)
- Energy sources (pumps for hydraulics and compressors for pneumatics).
- Actuators (linear or rotary).
- Valves (depending on the required operation).
- Tanks, filters, fittings, pipes and hoses.

The following section discusses the pumps used in hydraulic systems taking into account the procedure to select a pump and a few types of pumps used in different applications. Compressors of pneumatic systems are discussed in the following section. Linear and rotary types of actuators are analysed in the next section while different types of valves will be handled in the last section.

7.2 Pumps

Pump is the main source of energy in a hydraulic system. If there is no load on the pump, the main output of the pump is its flow rate of incompressible fluid. The mechanical energy given to the fluid by the pump can be transformed to hydraulic energy if a force is exerted by the load. In other words, the cumulation of the fluid flow leads to increasing the pressure of fluid in the system. Positive displacement pumps are used in hydraulic power for their capability of reaching high pressure that can reach more than 1000 bar.

7.2.1 Pump selection

Many parameters should be taken into account before selecting the type of pump:

- Maximum pressure in the system.
- Maximum required flow rate.
- Speed of the prime mover.
- Control technique.
- Fluid type and contamination level.
- Noise tolerance.
- Size and weight.
- Efficiency.
- Cost.
- Availabilty of the pump.

Taking into account the previous parameters, the following procedure can be followed to select a pump and hydraulic system successfully:

1. Select the type and size of the required actuator depending on the force exerted by the existing load.
2. Calculate the delivery flow rate needed from the pump to give a specific speed of flow.
3. Calculate the maximum pressure in the system depending on the force and area of the pipes and actuators.

4. Determine the speed of the pump and prime mover. This is related to the pump's delivery (displacement).

5. Determine the type of the pump that can give the previously calculated parameters.

6. Select the size of the tank and piping needed for the system. This includes the calculation of the pressure and head losses due to friction and head difference.

7. Estimate the cost of the system.

The following subsections will discuss a few types of pumps mostly used in the different applications.

7.2.2 Gear pump

Gear pump is one of the simplest designs of pumps. It consists of two gears in mesh where one of them is connected to the prime mover and rotates the second one by teeth contact. The gear mesh causes a high noise level in the pump. The pumping occurs by enclosing a small amount of fluid between the teeth and the case of the pump as shown in Figure 7.1. The torque on the gear shaft transfers the fluid from the low pressure part at the inlet to the high pressure side at the outlet of the pump causing a continuous flow that can reach up to 700 l/min. The pressure at the outlet depends on the force exerted by the load and can reach up to 150 bar. Gear pumps are very sensitive to particle contamination of the fluid where it is necessary to filter the fluid before the pump to avoid any erosion of the teeth and casing caused by friction due to dirt and partciles in the fluid. Any wear in the gear or housing leads to leakage of fluid and reduces the efficiency of the pump. Another type of gear pumps uses an internal gear in mesh with an external gear with smaller diameter. In this later case, a crescent is enclosed between the two gears to contain the fluid and transfer it from the inlet to the outlet. The delivery of this type of pumps can reach up to 750 l/min and the operation pressure ranges from 170 to 300 bar.

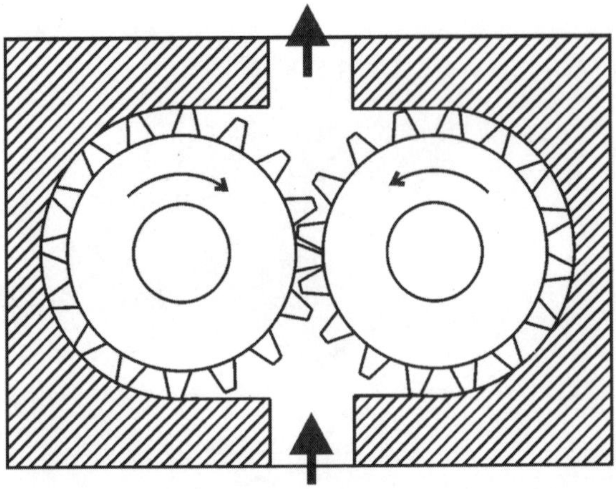

Figure 7.1: *Gear Pump*

7.2.3 Lobe pump

Another type of gear pumps is the lobe pump shown in Figure 7.2. Lobe pump consists of two gears in mesh where each gear is made of three teeth only. This type of pumps leads to a higher delivery due to the higher volume enclosed between the teeth and the casing but the reduction in the number of pulsations causes less continuity of flow compared to the normal gear pumps. This pump is less noisy than the previous one because each lobe is driven separately which reduces the contact between them reducing the noise level.

7.2.4 Gerotor pump

The gerotor pump shown in Figure 7.3 is made out of an internal gear in mesh with an external one that has one tooth less than the outer gear. The displacement of this pump depends on the volume enclosed between the two gears due to the difference of one tooth. All types of gear pumps have the problem of high leakage because of the necessity to have a clearance between the gears.

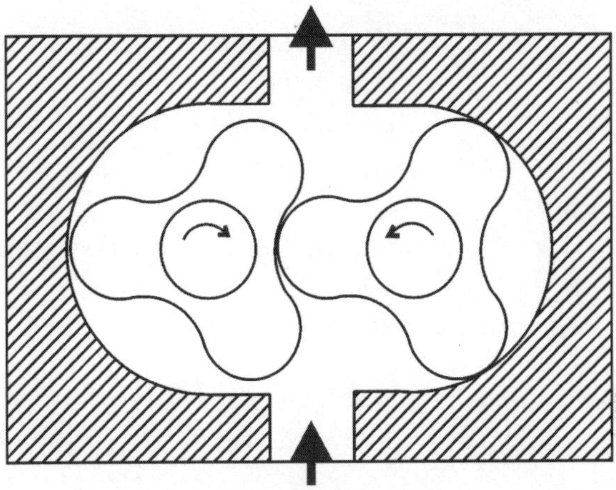

Figure 7.2: *lobe Pump*

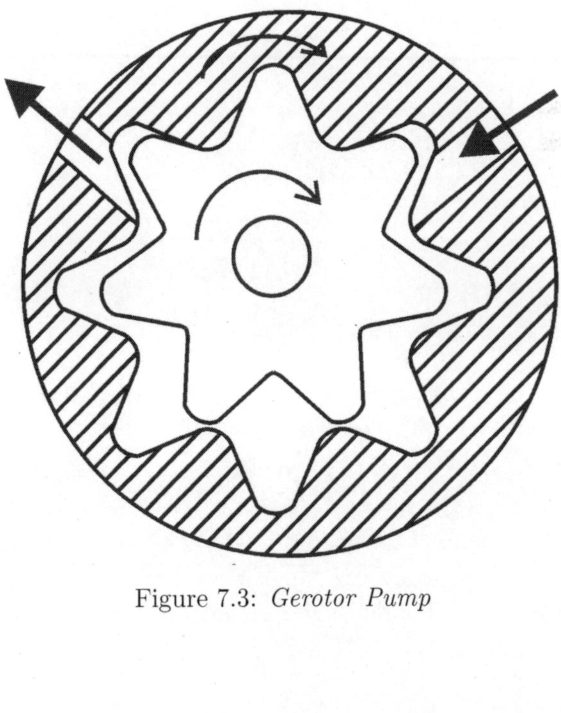

Figure 7.3: *Gerotor Pump*

7.2.5 Vane pump

Vane pump shown in Figure 7.4 consists of a rotor that contains radial slots and rotates inside a cam ring. Vanes are plates sliding radially inside the slots due to either the centrifugal force or pre-loaded springs under the vanes. Unlike the gear pumps, the problem of high leakage is solved in vane pumps because the vanes are pushed against the cam ring but this causes a high wear for the vanes. Eccentricity of the rotation in vane pump causes side force on the shaft. This problem is solved by using an elliptical cam ring with the rotor in the middle which is called balanced vane pump. The delivery of this type of pumps can reach up to 600 l/min and the operation pressure ranges from 100 to 170 bar.

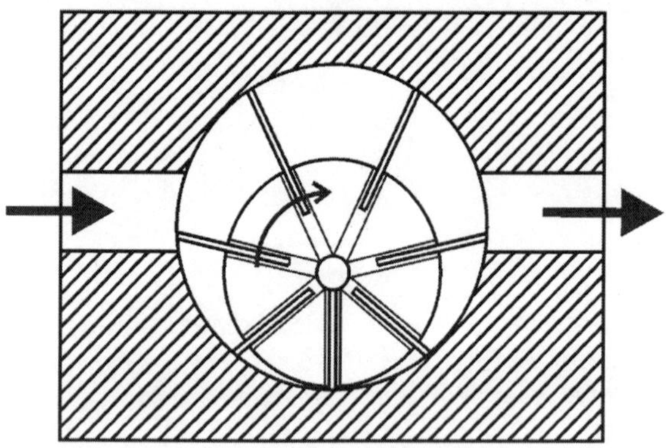

Figure 7.4: *Vane Pump*

7.2.6 Radial piston pump

The radial pump shown in Figure 7.5 consists of a ring housing and a concentric barrel containing the cylinders radially. A cam shaft rotates eccentrically inside the barrel to push the pistons radially outwards inside the cylinders. The pistons stay in contact with the cam by the effect of springs. The cam shape of the shaft's rotation causes the pistons to extend with different strokes depending on their position relative to the cam. This motion causes the cylinders to extract the fluid at the minor radius of the cam and deliver it at the major radius. The displacement of the pump depends on the cross sectional area of the pistons and the difference between the minor and major diameters of the cam. The delivery of this type of pumps can reach up to 1000 l/min and the operation pressure can reach up to 1500 bar.

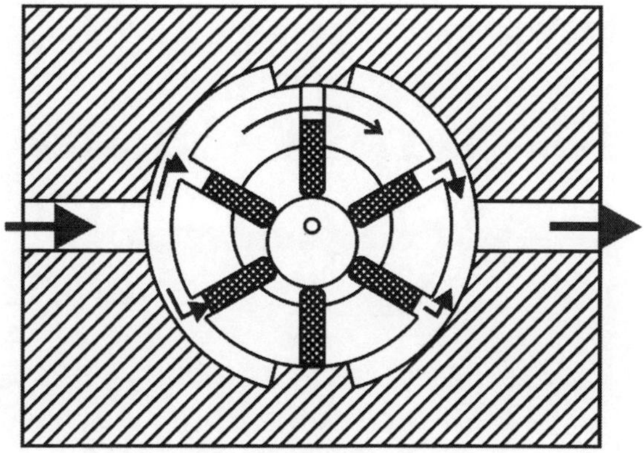

Figure 7.5: *Radial piston Pump*

7.2.7 Axial piston pump

Figure 7.6 shows an axial pump (swash plate design). The pistons are included axially in a rotating barrel, the rods of the pistons are connected to a swash plate inclined with a specific angle and connected to the barrel by a spherical or universal joint. The angle of inclination of the swash plate determines the displacement of the pump by changing the stroke of the different pistons. This type of pumps is very common in variable displacement pumps and very widely used in a wide spread range of applications. The delivery of this type of pumps can reach up to 3500 l/min and the operation pressure ranges from 200 to 350 bar. The same influence can be obtained by using the bent-axis axial pump shown in Figure 7.7

7.3 Compressors

The main component in pneumatic system is the compressor. Compressors are usually driven by electric motors to produce pressurized air and

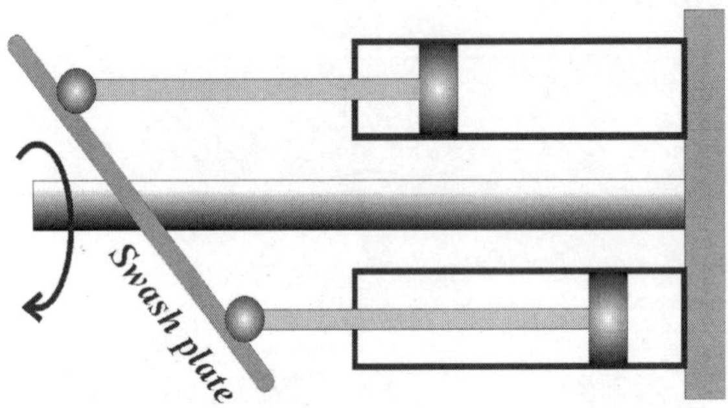

Figure 7.6: *Swash plate axial piston Pump*

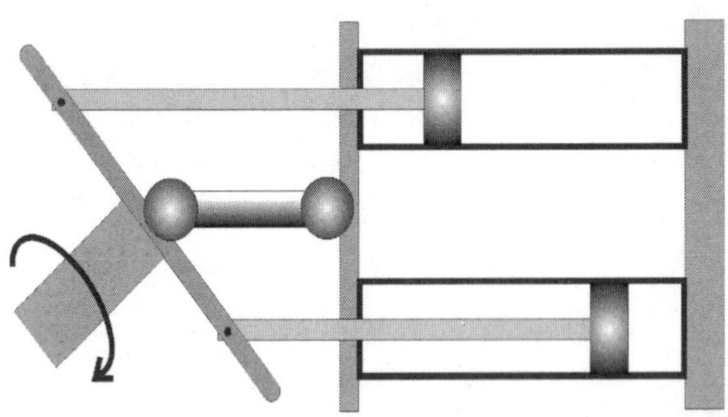

Figure 7.7: *Bent-Axis axial piston Pump*

store it in a closed reservoir to be used in exerting force and pressure on loads. Unlike hydraulic systems, pneumatic systems are open to the atmoshphere where compressed air is exhausetd out after being used. Compressing air leads to increase its temperature which explores the need to use aftercoolers to reduce its temperature. In the case of two-stage compressors, intercoolers are used between the stages. Filters and water traps are used to get rid of the water entraped in the compressed air. This water comes from the condensation of vapour existing as a humidity in the atmoshperic air. Positive displacement compressors are mostly used in pneumatic systems because they have the ability to produce higher pressure.

7.3.1 Piston compressors

The most commonly used compressor is the piston type shown in Figure 7.8. The basic principle of piston compressors is based on turning rotational motion into linear motion through a slider-crank mechanism where the rotational speed of the crank is transformed to a linear motion of the piston inside the cylinder. When the piston is pulled down, the inlet check valve opens and allows air to enter to the cylinder. Pushing the piston upwards in a return stroke, closes the inlet check valve and opens the outlet check valve forcing the air in the cylinder to flow out. A single-stage piston compressor can compress the air up to 15 bar.

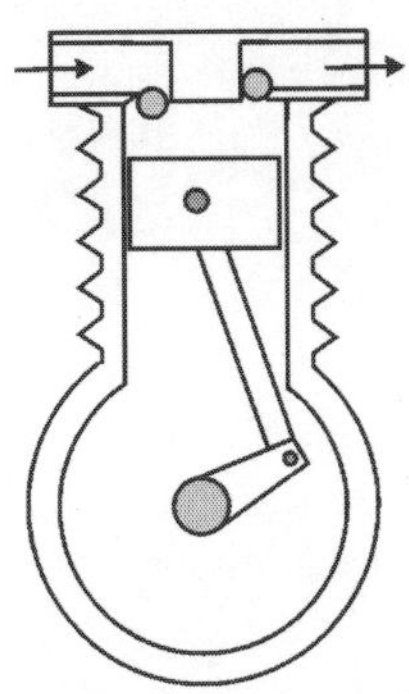

Figure 7.8: *Piston Compressor*

To get higher pressures that may reach up to 50 bar, it is necessary to use a multi-stage compressor. The two-stage compressor shown in Figure 7.9 is built of two piston compressors, the large one is used for low pressure compression while the smaller one compresses the same air to a higher pressure. An intercooler is used between them to reduce the air temperature before being compressed to the maximum pressure. Both stages are usually driven by the same motor.

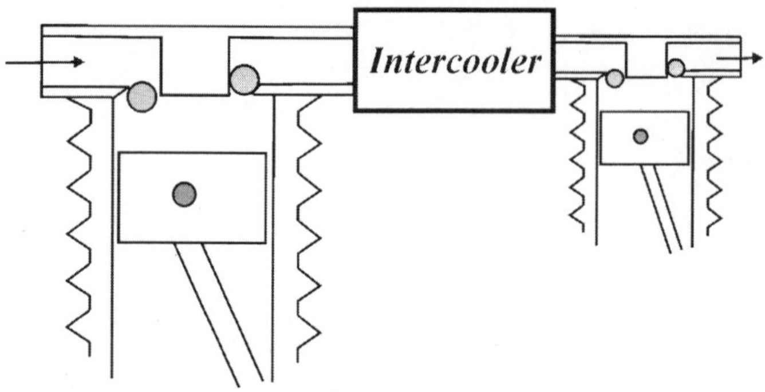

Figure 7.9: *Two-stage piston Compressor*

7.3.2 Diaphragm compressors

Piston compressors can cause a contamination to the outelt which makes a problem in food and medical industries. This problem is solved by separating air from moving parts by a flexible diaphragm as shown in the diaphragm compressor in Figure 7.10. Diaphragm compressors are often used in food, chemical and medical industries to avoid contamination from reaching to products.

7.3.3 Vane compressors

Vane compressors (Figure 7.11) have a similar design to vane pumps where the vanes can be loaded by springs or centrifugal force. This type of compressors cannot reach more than 3 bar pressure unless it is designed in two-stage where it can reach up to 10 bar.

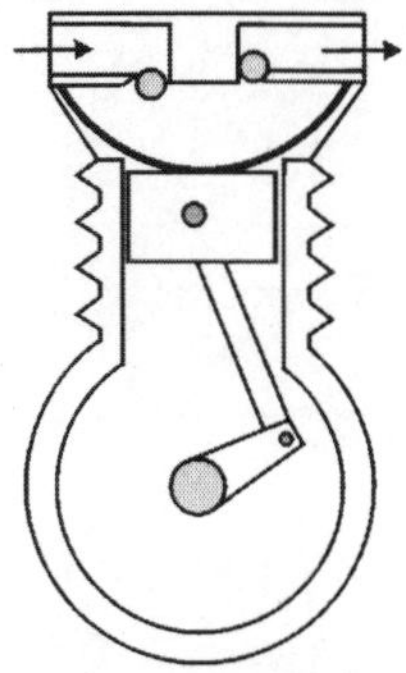

Figure 7.10: *Diaphragm Compressor*

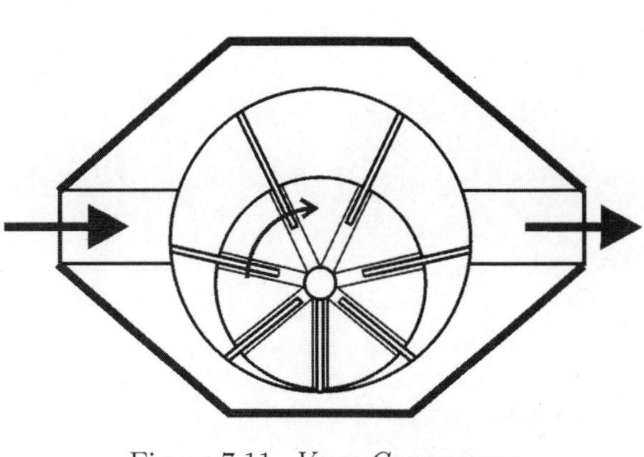

Figure 7.11: *Vane Compressor*

7.3.4 Lobe compressors

Another type of rotary compressors is the lobe compressor shown in Figure 7.12. Lobe compressors are used to produce high volume flow rates but with lower pressure 1 - 2 bar. Rotating the lobes leads to enclose the air between the lobes and the casing and transfer it towards the outlet.

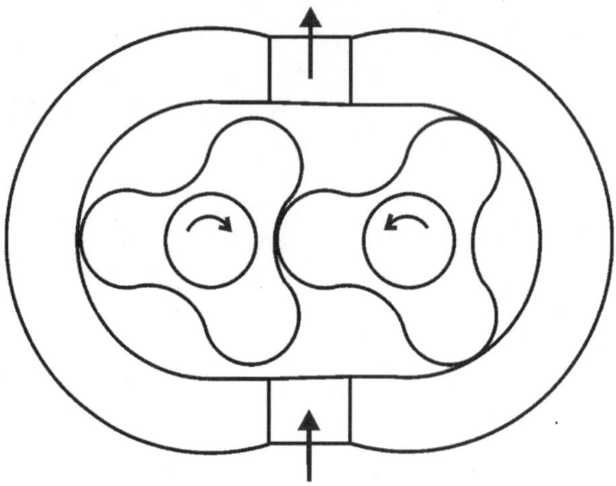

Figure 7.12: *Lobe Compressor*

7.3.5 Screw compressors

Screw compressor shown in Figure 7.13 is usually used in the applications where a high pressure (more than 20 bar) with a very low flow rate is required. High accuracy srews with very small clearance are used to enclose the air between the screws and the casing pushing it towards the outlet.

7.4 Actuators

Fluid power remains useless unless it is transferred to act on a specific load. This action can be done by an actuator. Actuators can be either linear or rotary depending on the required type of motion. A linear actuator consists of a cylindrical piston sliding axially inside a hollow cylinder by

Figure 7.13: *Screw Compressor*

the influence of the force exerted on it by the fluid pressure. The resulting froce is transmitted to the load by a rod connected to the piston. This force reads

$$F = PA \tag{7.1}$$

Where F is the induced force, P is the pressure of the compressed fluid and A is the inside cross sectional area of the piston. Rotary actuators have different shapes and designs like vane, gear and lobe types. The rotation torque is transmitted to the load by means of a rotating shaft connected to the rotating surface exposed to the fluid's pressure. This torque can be calculated by

$$T = PAR \tag{7.2}$$

Where T is the induced torque, P is the pressure of the compressed fluid, A is the area of surface exposed to the pressure in the rotating element and R is the distance between the center of the action of the pressure on the area A and the center of rotation of the rotating shaft. The following sections discuss different types of actuators.

7.4.1 Single-acting actuator

The simplest design of an actuator is the single-acting actuator shown in Figure 7.14. This actuator is used to exert force and displacement in the extension stroke while the retraction stroke is actuated by the return spring installed internally or externally on the actuator.

Figure 7.14: *Single-acting actuator*

7.4.2 Double-acting actuator

To obtain double sided controlled motion of the linear actuator, the double acting piston in Figure 7.15 is used. A double-acting actuator consists mainly of a piston, rod, cylinder and two caps. The internal surface of the cylinder barrel should have a smooth surface finish to reduce the friction effect caused by the contact between the piston and the cylinder. Piston seals are wound around the piston as rubber gaskets to avoid leakage of fluid through the clearance between the piston and the cylinder. End seals and wiper seals are other rubber gaskets installed between the rod and the cylinder to avoid leaking the fluid out of the actuator. Compressing the fluid into the actuator from the extend port leads to extend the piston with the desired displacement while the exerted force depends on the operational pressure. On the other hand, retraction is obtained by pushing the fluid into the retract port.

7.4.3 Telescopic actuator

Figure 7.16 depicts a two-stage telescopic piston used to have a longer actuation with a longer distance of the end rod. When oil is applied to the blank side of piston A prssure is applied to both sides of the piston exerting different forces on both sides due to different areas. The difference

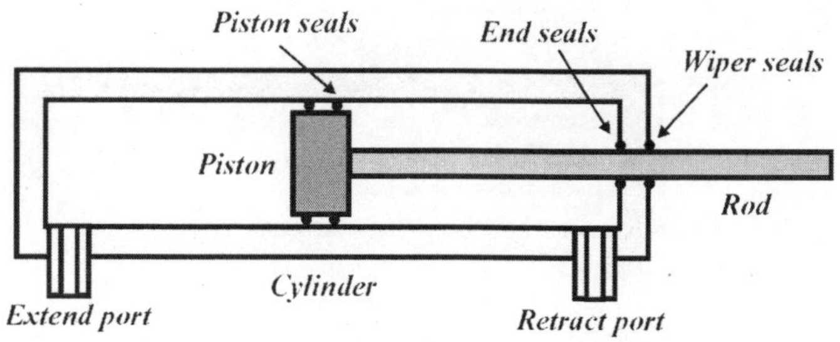

Figure 7.15: *Double-acting actuator*

of the force causes piston A to move to the right till reaching the full stroke then piston B begins to extend. To retract the piston, oil is applied to the port on the rod side of piston B retracting piston B first followed by piston A.

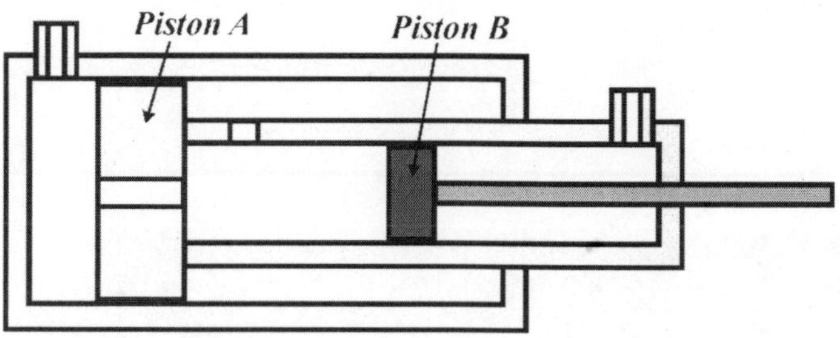

Figure 7.16: *Telescopic actuator*

7.4.4 Cushions of actuator

Cushioning is used to deccelerate the speed of extension or retraction of the piston near the end points of the stroke. Figure 7.17 shows a decceleration cushion added at the end of the retraction stroke of a piston. A plunger is installed at the blank side of the piston. This plunger enters inside a cylinder installed at the end cap of the cylinder reducing the flow of the exhausting fluid from the cylinder which is forced to exit through the needle valve slowing down the piston speed.

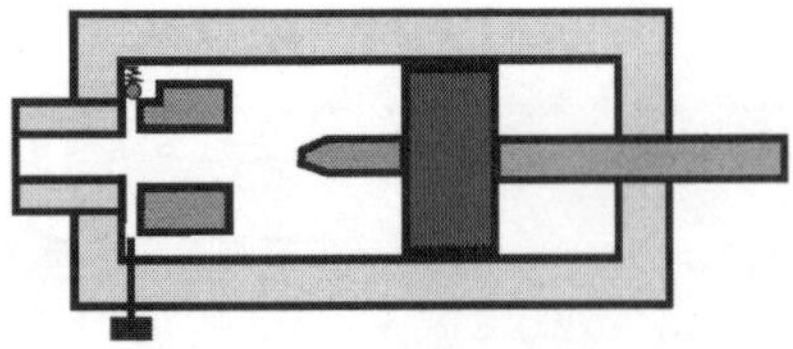

Figure 7.17: *Cushioning of an actuator*

7.4.5 Gear motor

Rotary actuators can have one of three designs; gear motors, vane motors or piston motors. Gear motor (shown in Figure 7.18) consists of two gears in mesh enclosed in a casing. The inlet port enters the fluid at high pressure to rotate the gears passing between the teeth and casing to be exhausted from the low pressure port. The direction of rotation is as shown in Figure 7.18. Gear motor suffers from leakage between the teeth and casing bacause of the need to have enough clearance.

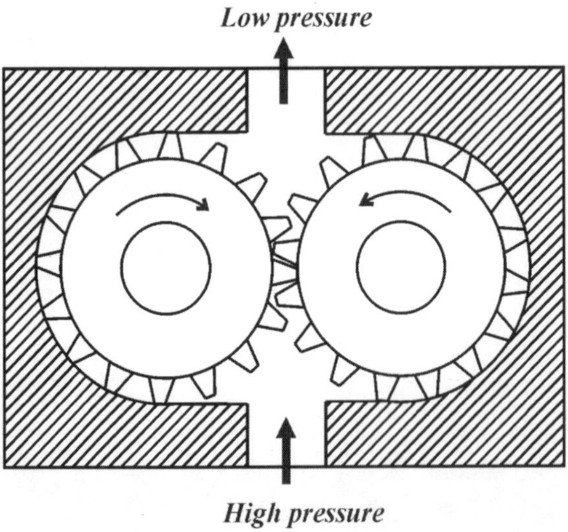

Figure 7.18: *Gear motor*

7.4.6 Vane motor

Vane motor shown in Figure 7.19 is similar to the design of the vane pump discussed earlier. Vane motor consists of several vanes sliding around a rotor inside a cam casing. Vane motors have less leakage effect than gear motors but the contact between the vanes and the casing is exposed to high friction effect. This type of actuators is usually used in low speed applications. Force unbalance can be solved by using dual design.

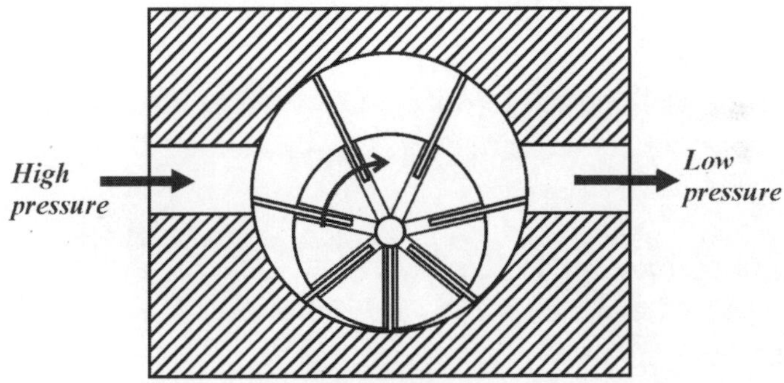

Figure 7.19: *Vane motor*

7.4.7 Piston motor

Higher torques and speeds can be achieved by using piston design actuators. Figure 7.20 shows a piston rotary actuator where compressed fluid extends the axial pistons leading to rotate the plate. Speeds are adjusted by changing the inclination angle of the bent-axis just like the piston pump.

7.5 Valves

Any hydraulic or pneumatic system needs different types of valves to direct and regulate the flowing fluid either oil or air before reaching to the application. These valvas can be divided into two main types:

- Infinite position valves: this type of valves can take any required position to close and open gradually determining the area of the

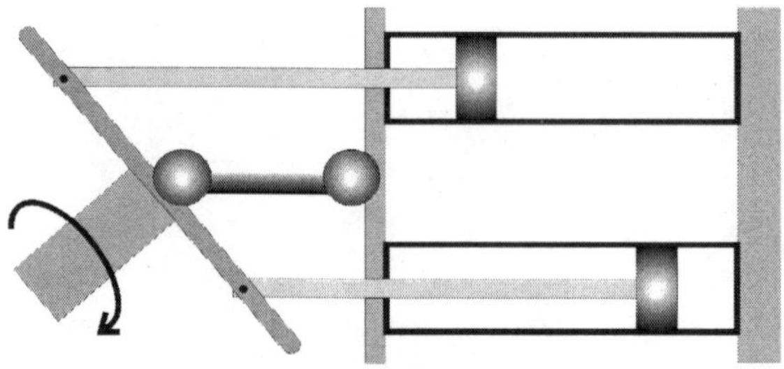

Figure 7.20: *Piston motor*

port (orifice) in order to control the flow rate that in turn regulates the speed of the actuator. Besides to the fact that it can stop at any position.

- Finite position valves: these valves are also called ON/OFF valves that serve to either fully open or fully close to pass or to stop the fluid flow.

The conceptual design of the different types of valves is almost the same for hydraulic and pneumatic applications. The only difference is the need for stronger materials and seals in the case of hydraulics because they need to support higher pressures. The following sections discuss different general designs of valve types.

7.5.1 Check valve

Check valves, or called non-return valves are used to keep the direction of flow in one sense of direction preventing the fluid from flowing back in the opposite direction. Figure 7.21 shows a simple design of the check valve where a ball mounted on a prestressed spring is used to close the orifice in one direction. When the force induced on the ball's surface by the fluid pressure exceeds the spring force, the ball is pushed down allowing the fluid to flow through the valve. The arrows on the drawing determine the direction of flow that cannot be reversed unless the ball is pushed against the spring by another mean like a mechanical handle, pliot pressure or a solenoid.

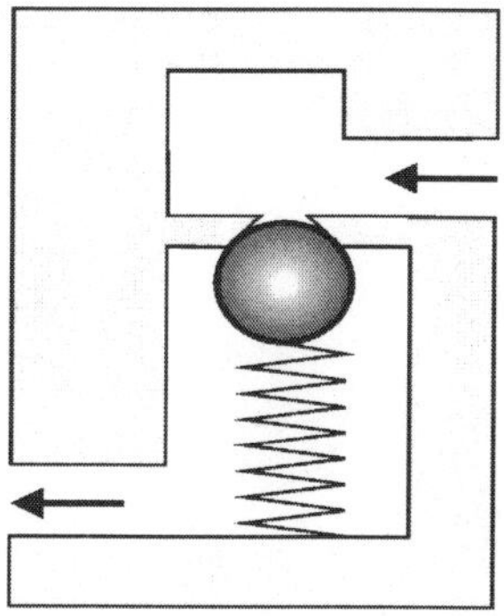

Figure 7.21: *Check valve*

7.5.2 Poppet valve

Poppet valve shown in Figure 7.22 is similar to the the check valve discussed in the previous section except that the poppet valve is operated by a manual handle that serves to push the ball againts the spring force to open the orifice and allow the fluid to pass freely in the direction shown in the figure. In the opposite direction, the valves behaves like a normal check valve.

7.5.3 Spool valve

Figure 7.23 shows a directional control spool valve. Spool valves are the most commonly used valves to control the direction of flow. Spool valve consists mainly of a spool rod with cylindrical lands allowed to slide inside an envelope vented to the main inlets and outlets of the valve. When the spool is pushed to the right, pressurised fluid coming from the pump flows through port B to the application while port A is open to the tank in this case to drain the fluid existing in the piston back to the tank. Pushing the spool to the left reverses the operation. Spool can be operated either manually or automatically using a pilot pressure or a solenoid. A third action can be achieved by keeping the spool in a center position by two springs mounted on both sides of the valve.

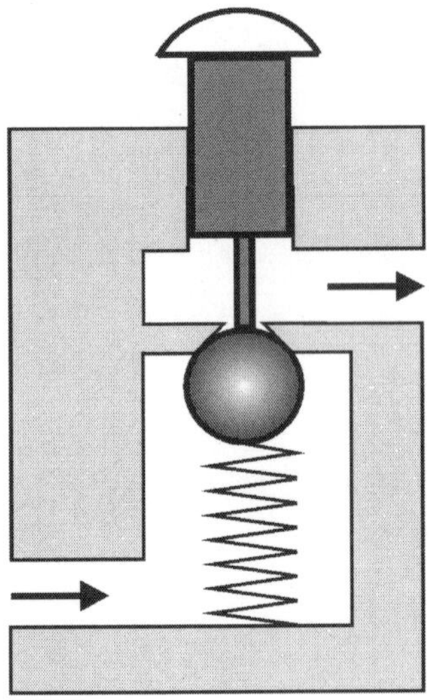

Figure 7.22: *Poppet valve*

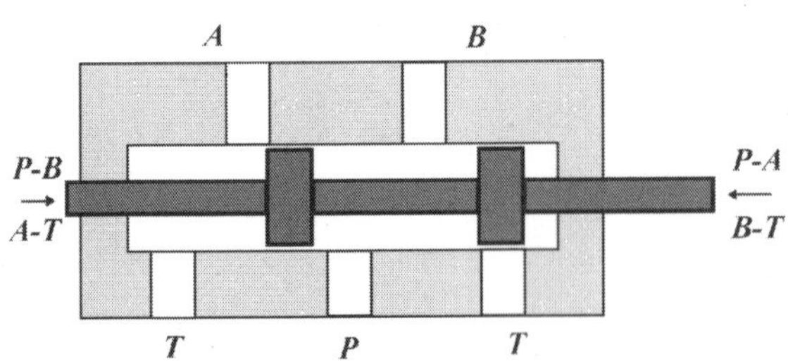

Figure 7.23: *Directional control spool valve*

7.5.4 Rotary valve

The direction of fluid flow can be controlled by rotary valves as the design shown in Figure 7.24. The outer cylindrical casing containing the inlet and outlet ports is fixed. The inner rotating rotor consists of two perpendicular channels used to connect specific ports together when rotated. In the shown position, the pressure port is connected to port B and port A is connected to the tank. Other positions can be obtained by rotating the internal rotor either clockwise or counter-clockwise.

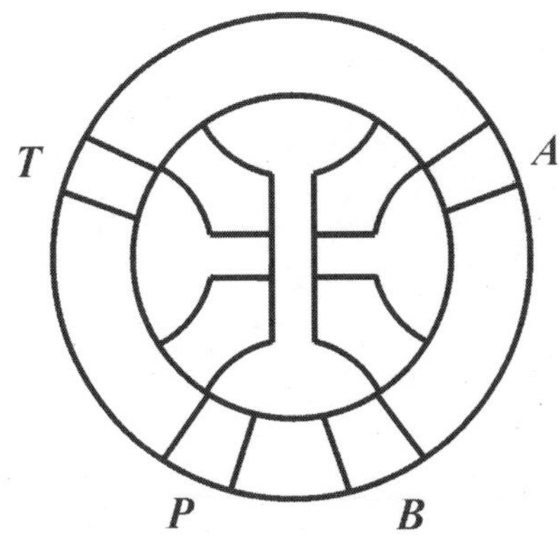

Figure 7.24: *Rotary valve*

7.5.5 Shuttle valve

Shuttle valve or double check valve shown in Figure 7.25 is used to keep a constant pressure in the system when the flow comes from either of two inlets X or Y. The valve consists of a ball enclosed in an envelope. When the fluid enters from inlet X, the ball is pushed to close inlet Y allowing the fluid to flow through the valve to port A. The same action occurs when the flow comes from inlet Y closing X and allowing fluid to flow from Y to A. Shuttle valves are rarely found in hydraulic systems, rather they are commonly used in pneumatic circuits. In some designs, ball of the shuttle valve can be loaded by a spring to one side and can be pushed to the other side by the influence of fluid pressure.

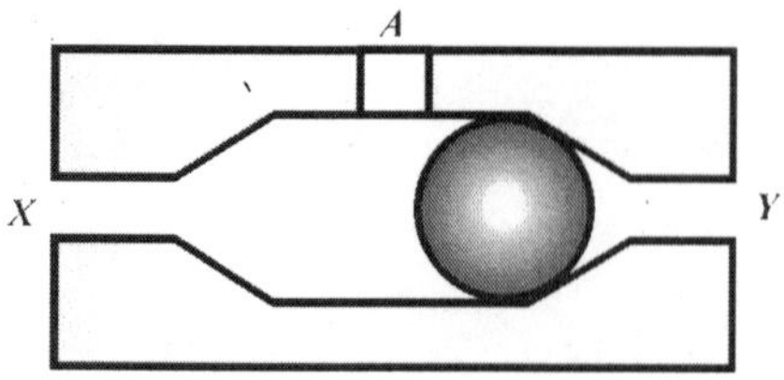

Figure 7.25: *Shuttle valve*

7.5.6 Sequence valve

Sequence valve is a combination of a directional control valve and a pressure relief valve. Figure 7.26 shows a sequence valve that consists of a directional control spool valve connected to two pressure relief valves. When the spool moves to allow the fluid to extend piston 1 the pressure relief valve remains closed till piston 1 reaches its end stroke and thus the pressure increases in that line, at this moment, the relief valve opens allowing the fluid to flow to extend piston 2 in a sequence form. The same operation occurs in the retraction stroke.

7.5.7 Pilot operated valve

Figure 7.27 shows a pilot operated check valve. The pilot fluid flows to push the small piston connected to a pushing rod that serves to push the ball against the spring opening the main port to allow the operation fluid to pass to the application. The main advantage of the pilot operation of the check valve is that it enables the valve to allow the reversed flow of fluid when actuated. Normally, the pressure of the pilot line does not exceed ten percent of the operation pressure.

7.5.8 Time delay valve

Figure 7.28 shows a time delay valve that can be used to delay the actuation of the valve delaying the operation with a time depending on the adjustment of the valve. Time delay valve contains a reservoir in the way of the incoming pilot fluid besides to a needle valve at the inlet. Delay of the operation is achieved by changing the orifice of the needle valve where

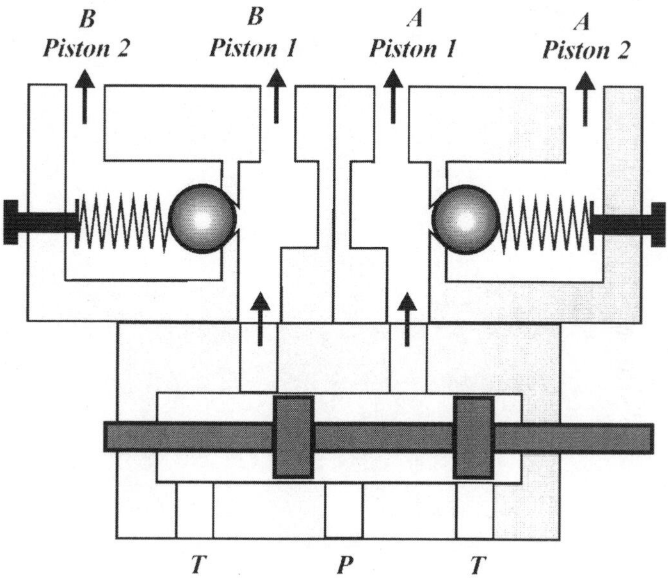

Figure 7.26: *Sequence valve*

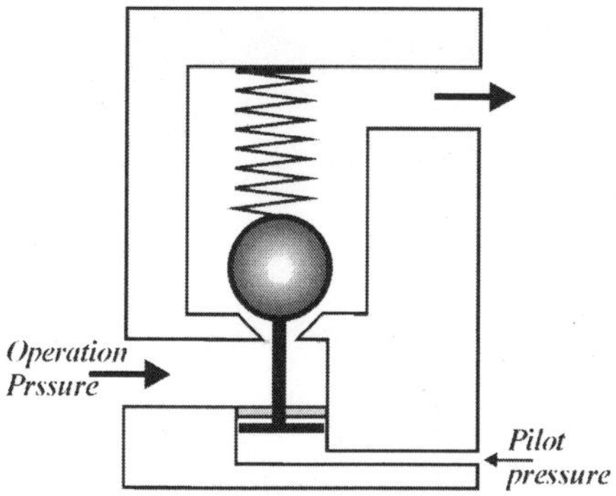

Figure 7.27: *Pilot opearted check valve*

filling the reservoir takes a time before reaching the required pressure to push the ball of the check valve.

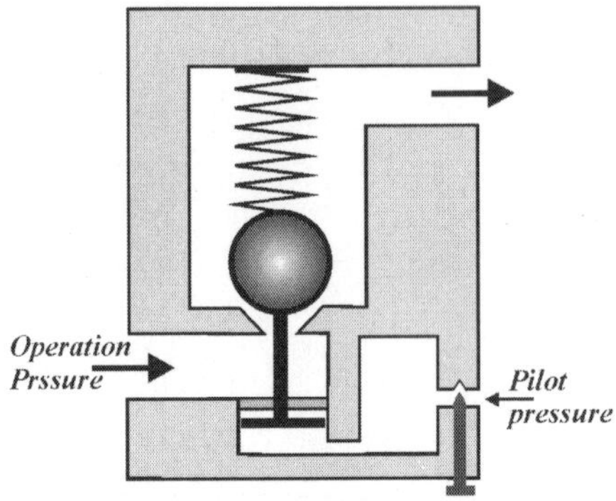

Figure 7.28: *Time delay check valve*

7.5.9 Modular valve

Modular valve is a way to connect valves to the inlets and outlets. This is done by having a base containing internal pipes and channels to connect between the different ports of the valves while several valves are connected to this base. Normally, the inlet to the modular valves is common while the oulets are different to reach to the different applications.

7.6 References

[1] A. Parr, *Hydraulics and Pneumatics, A technician's and Engineer's Guide*, Newnes, 1991.

[2] A. Esposito, *Fluid Power with Applications*, Prentice Hall, 2003.

Chapter 8

Fluid power applications

8.1 Introduction

Fluid power solutions have the greatest contribution in the modern industrial revolution. It is worth mentioning that there is a wide variety of applications for fluid power systems. Most of the heavy duty equipment working in the industry sector depend mainly on either hydraulic or pneumatic principle where the main power source in these systems come from oil in hydraulic systems and from air in pneumatic solutions. The early priciples of control depended on using mechanical techniques. Nowadays, most of the fluid power systems include electrical and digital control techniques which facilitates the use and implementation and gives higher performance to the power systems. The next sections discuss some hydraulic and pneumatic instruments used in industrial applications.

8.2 Hydraulic jack

The simplest hydraulic system is expressed in the hydraulic jack shown in Figure 8.1. This lifting jack is used to lift trucks and vehicles during maintenance process. It consists of a linear actuator supplied by a manual pump. The manual pump is a reciprocating piston operated by a manual handle to deliver the oil into the lifting piston. The manual pump is sometimes replaced by pneumatic compressed air that serves to exert force on the oil (air over oil system).

Figure 8.1: *Hydraulic jack (courtesy of Torin)*

8.3 Hydraulic loader

Figure 8.2 shows a picture of a hydraulic loader excavator vehicle used in dig and fill operations of construction. The whole system is driven by an internal combustion engine that provides the main pump with the required rotational speed and torque. The main hydraulic pump is a positive displacement pump (piston, vane or gear). The pump's oil delivery is accumulated in the system leading to increase the pressure. To actuate the loader, the compressed oil is transferred to the linear actuators through directional and flow control valves.

The loader is moved in two degrees of freedom; where two hydraulic pistons in parallel are used to lift it up while other two parallel pistons are used to rotate the bucket in a tilting motion. Every couple of pistons is influenced by the same load because they are attached to the same rigid body. Thus, the two pistons can be connected in parallel to double the force in keeping a synchronized motion.

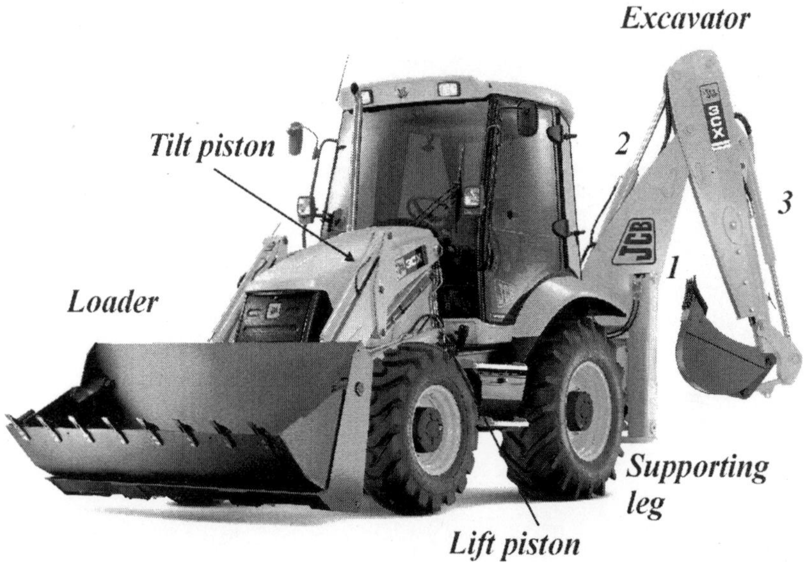

Figure 8.2: *Loader excavator machine (courtesy of JCB)*

8.4 Backhoe excavator

The backhoe excavator is used in construction works for digging in soil and aggregate. Some designs have only excavators driven by the Internal Combustion (IC) engines where the pump is used to supply and move the different motions of the backhoe shown in Figure 8.3.

An excavator has four degrees of freedom where two linear actuators are used to rotate the two arms of the boom, a third piston is used to rotate the end bucket to be filled with soil. The fourth motion here is for the rotational steering motion to rotate the whole system to both sides. The steering motion can be done either by using a rotary hydraulic motor or by linear actuators connected to racks that serve to rotate a pinion connected to the main column of the boom.

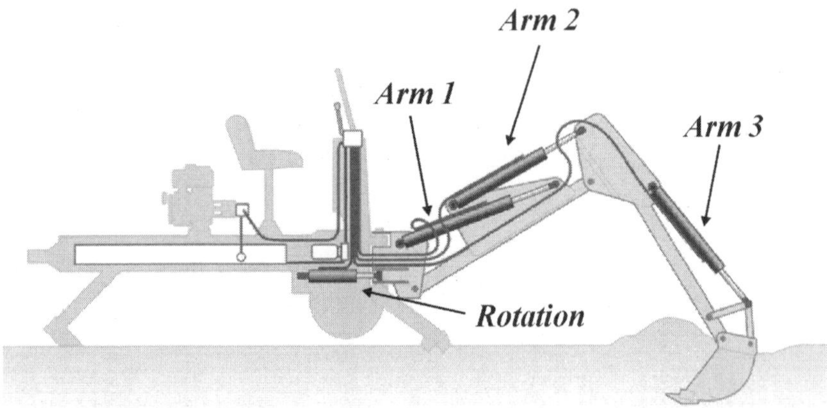

Figure 8.3: *Backhoe design and operation*

8.5 Concrete mixer

Concrete mixer shown in Figure 8.4 consists of a specially designed barrel able to rotate around its center. The rotation is achieved by rotating the main shaft centerd in the front side of the barrel while the back side has a circular ring supported by two steel rollers. Spiral blades are welded on the internal surface of the barrel to mix the concrete when rotated clockwise and to deliver it out from the back hole when rotated counter clockwise.

Figure 8.5 shows a schematic design of a mixer where the main rotational power is provided by a hydrostatic transmission system consisting of an axial piston reciprocating pump and a hydraulic motor. The pump can be installed directly to the flywheel of an IC engine or by taking power from the vehicle's gearbox through a power take-off instrument (PTO). Most of the hydraulic power is transformed into torque here because the rotational speed of the mixer does not exceed 20 rpm. The rotational speed of the mixer is controlled by both; increasing the rotation speed of the IC motor and changing the angle of the swash plate of the pump.

Figure 8.4: *Concrete mixer (courtesy of Liebherr)*

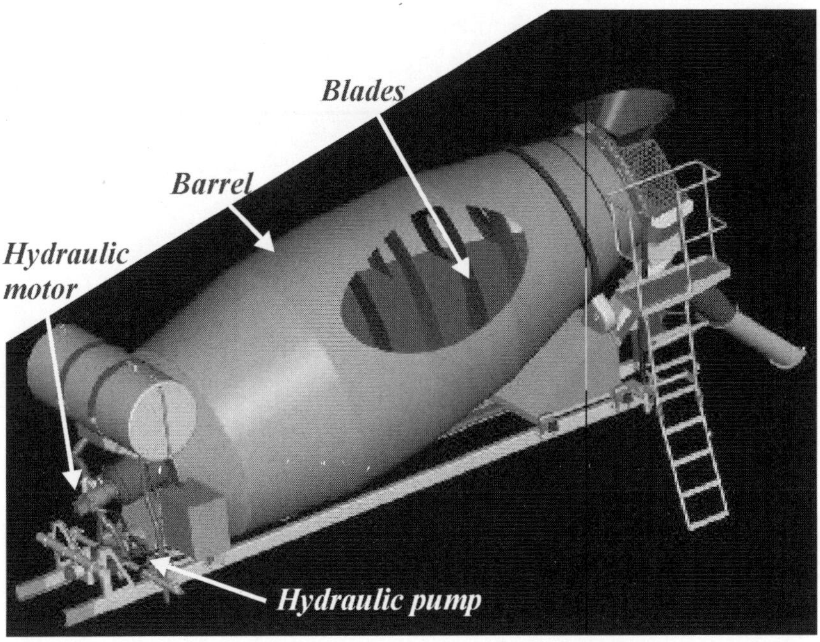

Figure 8.5: *Concrete mixer (schematic)*

8.6 Concrete pump

Concrete pump is considered one of the most important instruments used in construction works. The design of this machine depends mainly on the application requirements where it can be installed on a truck as shown in Figure 8.6 or it can be fixed on the construction site in the case of huge and high buildings. The operating system of this pump can be divided into three parts:

- Supporting legs.
- Extension boom.
- Pumping system.

To avoid exerting load on the body of the truck, four supporting legs are connected to the chassis of the pump. When these legs are opened, they make the shape of X letter which enables them to react against the moment coming from the boom's weight. Each leg contains two hydraulic pistons; one for extending the leg aside and the other is for lifting the whole system up. It is advised to extend the legs to the maximum distance away from the trcuk and to adjust their height till reaching a horizontally balanced situation for the pump in all directions. The ready mixed concrete is pumped through pipes to reach to the construction site. These pipes are installed and fixed to a long boom that can reach from 18 to 72 meters according to the required distance. The boom can be divided into three to six foldable portions depending on the design. Each portion of the boom is hinged at the end of the previous portion and rotated by a hydraulic linear actuator supported on the previous one. The rotational motion or steering degree of freedom is operated either by using a rotary hydraulic motor or by a linear actuator with rack and pinion mechanism.

The main system used to compress the concrete into the pipes is the pumping system shown in Figure 8.7. Pumping system consists of two concrete pistons operated by two hydraulic linear actuators and a rock connector rotated by a reversing piston. The two pumping pistons are

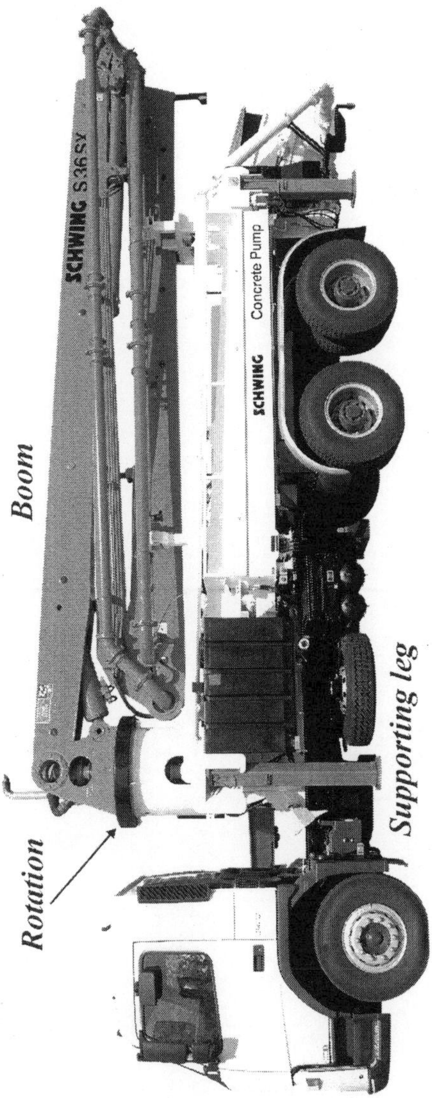

Figure 8.6: *Concrete pump (Courtesy of Schwing)*

connected in reciprocating manner where the extension stroke of one of them begins at the end of the retraction stroke of the other. The rock connector takes the shape of S or elephant trunc to transmit the concrete from the pumping piston to the transmission pipe. This rock is rotated by the reversing piston to be aligned with the pumping pistons in a sequence manner. This rotational motion is synchronized with the stroke of the pumping pistons. Limiting the strokes of the pumping actuators and synchronizing them with the reversing rock can be controlled in one of two ways:

- Mechanical control using pressure relief valves that open and close according to the pressure in the system.

- Electrical control using proximity sensors, relays and solenoids.

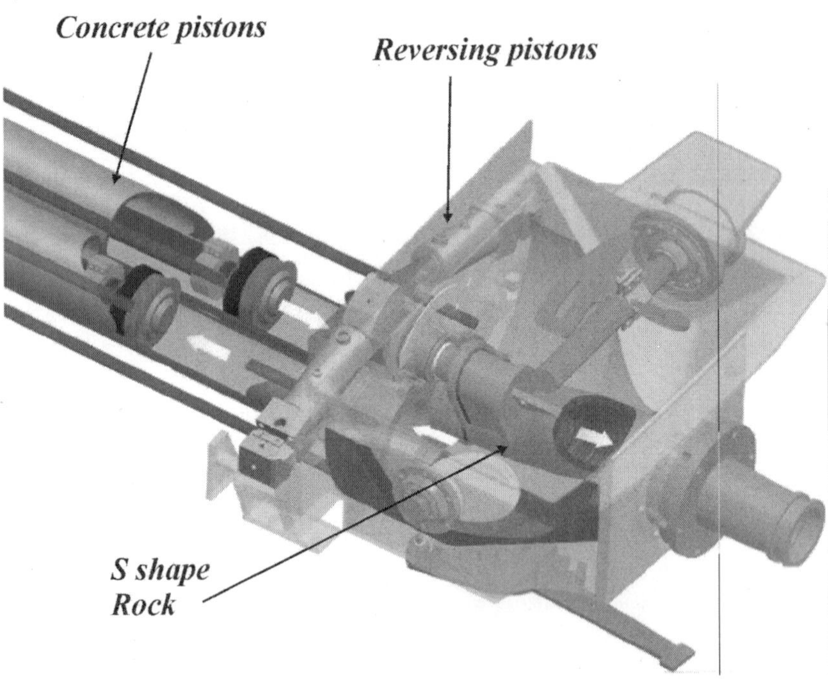

Figure 8.7: *Concrete pumping system*

8.7 Fork lift

Fork lift is a device used to lift loads and containers and transfer them from point to point. Figure 8.8 shows a schematic drawing of a typical fork lift. It requires two degrees of freedom to complete the lifting operation; lifting up and tilting back. These two motions are operated by two hydraulic pistons. The lifting piston must have a long stroke to enable raising the load to a high place while the tilting piston needs a very short stroke to tilt the load backwards by a maximum angle of 10 degrees to avoid droping it during motion. Sprokets and chains are sometimes used to extend the stroke of the lifting piston. The advance motion is done by moving the truck back and forth. This motion is sometimes operated by a hydraulic rotary motor connected to the wheels.

8.8 Hydraulic crane

Figure 8.9 shows a telescopic hydraulic crane used to lift and position loads to buildings and high places. Cranes are designed according to the required applications specifying the needed length and lifted weight. Apart from the extension telescopic motion, a crane has two main motions; lifting and rotation. Lifting motion can be operated by a linear hydraulic piston and the rotational motion can be done either by a hydraulic rotary motor or a linear actuator with rack and pinion mechanism. The extension of crane's length is done by a telescopic hydraulic actuator. Cranes need supporting legs to react againts exerted moment of the boom and load.

8.9 Metal scrap press

Metal scrap collected for recycling processes takes a very large volume and space. To reduce the required space, presses like the one shown in Figure 8.10 are used to press the scrap together in reduced size cubes. Scrap press consists of a U shaped chambre with a front covering gate. The upper side is functioned by a hydraulic piston to press the scrap from one side and the back side is functioned by another hydraulic piston to

Figure 8.8: *Hydraulic fork lift (courtesy of Montacarishi)*

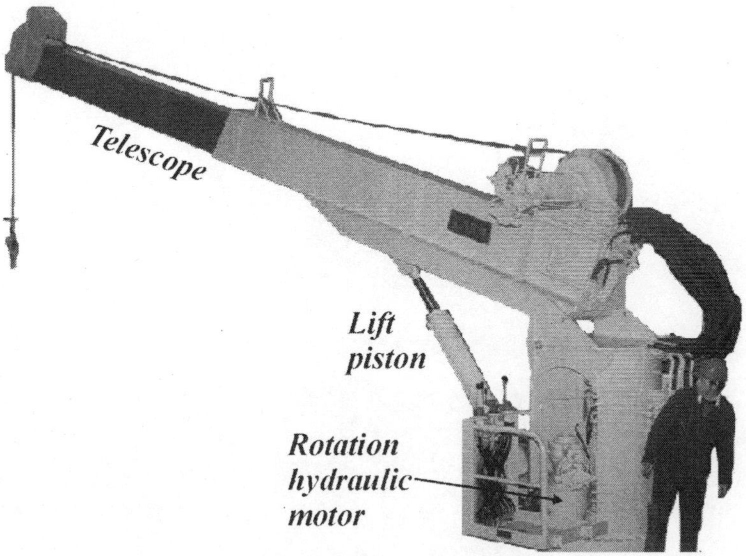

Figure 8.9: *Telescopic hydraulic crane (courtesy of Karmoy Winches)*

make the final pressing motion converting the scrap into a small cube. Eventually, the front gate is opened and the back press piston is extended to spill the cube out of the press.

8.10 Leakage testing machine

Production of water plastic bottles by blowing techniques is accompanied by leakage problems according to defects and errors done either by workers or by the machine itself. This leads to the need for a leakage testing machine like the one shown in Figure 8.11. A leakage tester consists of a testing head, a conveyor belt and an ejector. The testing head contains a pipe to blow compressed air and another pipe connected to a pressure gauge. The head is pushed down to cover the bottle by means of a pneumatic piston. During the testing operation, the blowing pipe blows compressed air inside the bottle till reaching a specific pressure and the pressure gauge measures the pressure value inside the bottle for a few

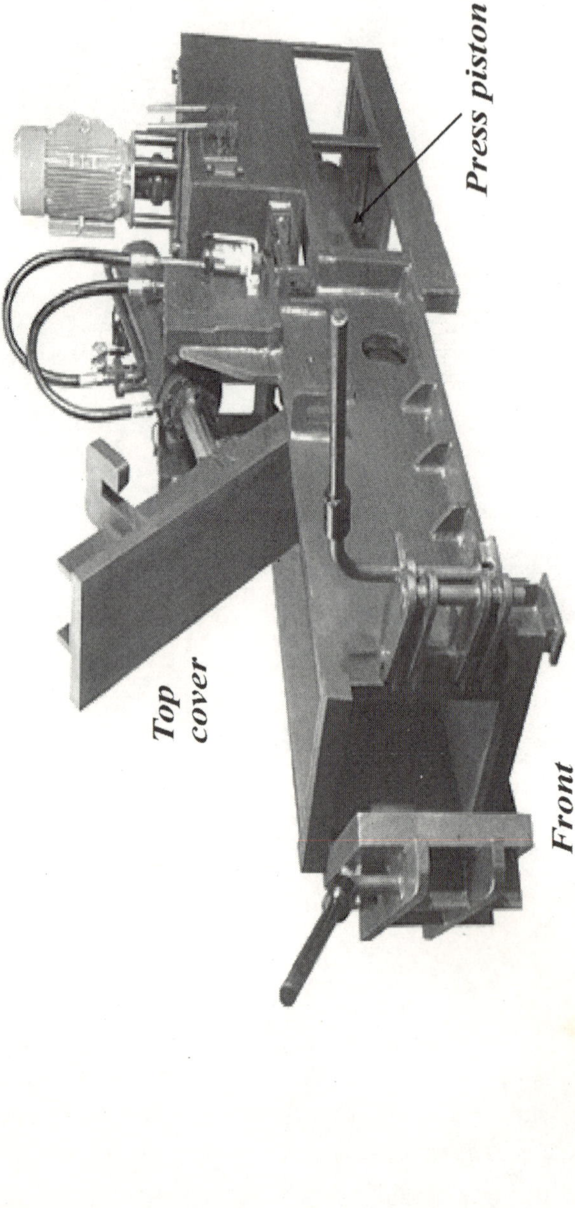

Figure 8.10: *Hydraulic metal scrap press (courtesy of Advanced Hydrau-Tech PVT)*

minutes. If the pressure stays at the same level, the bottle succeeds and passes. Otherwise, if the pressure descends down, this means having a leakage and the bottle fails the test. Failure of the bottle sends a signal to the ejector spring loaded, single acting pneumatic piston to extend and eject the defected bottle away from the line. The conveyor belt moves to approach the next bottle and positions it under the testing head.

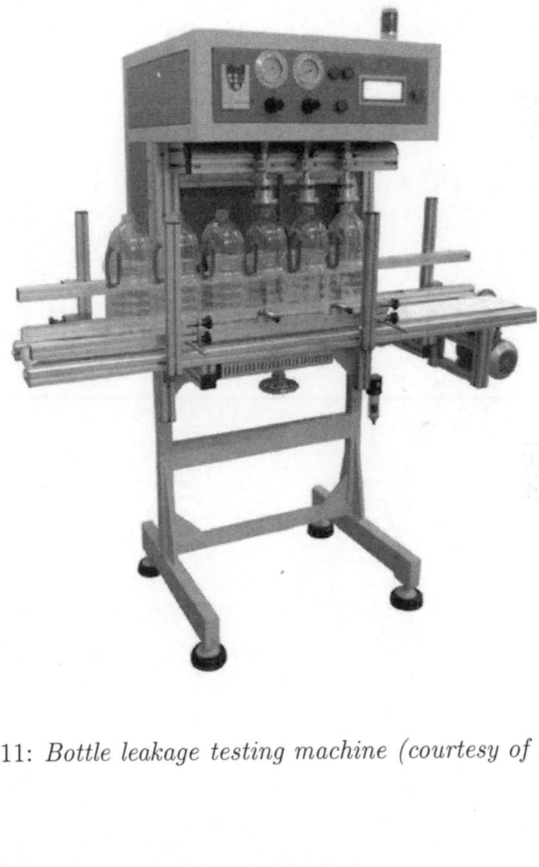

Figure 8.11: *Bottle leakage testing machine (courtesy of Piotech)*

8.11 Tyre changing machine

Flat tyres are changed or fixed using tyre changing machine as shown in Figure 8.12. Tyre changing machine contains three main systems:

- Detachment mechanism.
- Clamping table.
- Rolling out head.

The detachment mechanism is a blade moved by a pneumatic piston and used to detach the tyre from the metallic wheel on both sides. The clamping table contains four movable jaws actuated by two pneumatic pistons. The table is rotated by an electric or pneumatic motor for the purpose of tyre rolling out. The tyre is rolled out during the rotation of the table using the rolling out head that can be moved manually up and down.

8.12 Pneumatic industrial tools

Rotating hand-held tools can be operated using pneumatic power like the ones shown in Figure 8.13. The cutting tools in these machines are connected to a main rotor that acts as a turbine or a vane in the path of the compressed air. Different speeds and actions can be achieved using air pressure in these pneumatic cutting and industrial tools.

Figure 8.12: *Tyre changing machine (courtesy of Corghi)*

Figure 8.13: *Different pneumatic tools (courtesy of Industrial Hardware)*

Chapter 9

Maintenance of fluid power systems

9.1 Introduction

Prevention is better than cure. This is the best sentence to begin with before talking about maintenance of mechanical systems in general especially fluid power systems. Most of working companies spend a lot of money training their maintenance personel to troubleshoot a hydraulic system. Although, they should focus on preventing system failure to spend less time and money on troubleshooting a hydraulic or pneumatic system. In the following sections, a detailed discussion about maintenance and troubleshooting of both hydraulic and pneumatic systems will be presented.

9.2 Maintenance of hydraulic systems

9.2.1 Maintenance categories

Maintanance of hydraulic systems can be focused in the following categories:

1. *High temperature of oil:* High temperature influences deeply the properties of oil, like viscosity which reduces the efficiency of the system. To prevent that, a fluid temperature alarm should be installed in the system and all high temperature indications must be investigated and rectified immediately.

   Excess heat of oil can be caused by the following:

- Wear of the pump.
- Oil contamination.
- The air flow through the oil cooler (heat exchanger) is too low .
- The oil level is low.

2. *Cavitation:* Occurs when the volume of fluid demanded by any part of a hydraulic circuit exceeds the volume of fluid being supplied.

3. *Damage of seals and rings:* Cylinders have seals and rings that can be damaged by excess pressure and contaminants in the fluid. to avoid that make sure that the hydraulic fluid is clean. A cylinder is designed to take loads along its axis only. Side loads can decrease cylinder's life by causing excess wear on seals and rod.

4. *Contamination:* Contamination have two types:

   - Particulates: Such as dirt, sand, metal or rubber wear particles that come from internal wear during maintenance, attachment changes and machine operation.
   - Chemicals: They come from oil, water and air.

   Fluid contamination damages hydraulic system in two ways: the first is that it reduces system efficiency. Efficiency losses usually occur slowly and can reach (20 %) before the operator detects a loss in performance. These invisible efficiency losses also can increase fuel consumption. Contamination also accelerates component wear, where (75-85 %) of hydraulic pump, motor, cylinder and valve failures can be traced to contamination. Contamination can be avoided by the following ways:

   - Oil storage and transfer by using transfer filter cart.
   - Change filters carefully.
   - Carefully hose assembly and storage.
   - Avoiding mistakes in changing filters.

9.2.2 Recommended Maintenance technique

The following steps can be followed to apply the best maintenance technique:

(a) Perform daily inspections.

(b) Writing maintenance reports.

(c) Inspection of valves and other parts.

(d) Watch temperature and pressure gauges.

(e) Keep hydraulic tank filled.

To ensure a long life and reliability of the hydraulic systems, the most important maintenance recommendations include:

1. Keep hydraulic fluids cool. The bulk oil temperature at the exterior of the reservoir should not exceed 60^oC and the exterior of all components must be kept clean to ensure that no hot spots develop as a result of accumulated dust and dirt.

2. Keep hydraulic fluids dry. Water content generally should never exceed 1000 ppm (0.1%) in hydraulic systems using mineral base or synthetic fluids.

3. Repair fluid leaks immediately. If oil can escape, dirt, dust and air can re-enter the system. Keep in mind that an external leak of one drop of oil per second is equal to 1600 litres in a 12 month period.

4. Keep hydraulic fluid clean. It is known that 75 to 80% of hydraulic component failures are caused by fluid contamination with dirt, water, wear particles and other foreign material.

5. Establish an effective oil analysis program. The fluid used in a hydraulic system is a critical component of that system and its condition should be monitored as part of an effective maintenance and reliability program.

9.2.3 Problems, possible causes and remedies

The following tables show some problems occur in hydraulic systems with their possible causes and ways of solution to solve these problems.
Problem 1: Pump not pumping (Table 9.1)

Table 9.1: *Problem: Pump not pumping*

Possible cause	Remedy
Pump rotates in wrong direction	Reverse direction of rotation
Clogged intake	Check pipe from tank to pump
Low oil level	Fill to adequate level
Air leak intake	Check noise of pump
Pump speed too low	Check driving motor
Oil too heavy	Check viscosity and replace oil

Problem 2: Noisy pump (Table 9.2)

Table 9.2: *Problem: Noisy pump*

Possible cause	**Remedy**
Air leaking into system	Fill oil reservoir and check intake pipe
Air bubbles in intake oil	Check level of intake pipe
Pump cavitation	Look at cavitation problem next section
Loose pump parts	Tighten parts and check gaskets
Stuck pump vanes and pistons	Check for metallic contaminants
Dirty filter or strainer	Clean or replace
Small size filter	Check filter size
Pump running too fast	Determine appropriate speed
Return line above fluid level	Extend return line

Problem 3: Pump cavitation (Table 9.3)

Table 9.3: *Problem: Pump cavitation*

Possible cause	**Remedy**
Clogged or too small strainer	Clean or renew
Bore of suction line too small	Fit larger bore pipes
Too many bends in suction line	Modify pipe layout
Suction line too long	Reduce length or fit larger bore pipes
Fluid too cold	Heat fluid to recommended temperature
Unsuitable fluid	Replace with correct fluid
Air breather blocked	Clean or replace element
Restriction in suction line	Open or modify valves
Pump running too fast	Reduce to recommended speed
Pump mounted above oil level	Modify pump installation

Problem 4: System overheat (Table 9.4)

Table 9.4: *Problem: System overheat*

Possible cause	Remedy
Oil viscosity too high	Replace with lower viscosity
Internal leakage	Check for wear and loose parts
High discharge pressure	Proper setting of relief valve
High friction	Check loose pump parts
Clogged oil cooler	Clean cooler
Low oil level	Fill to proper oil level

Problem 5: Low or no pressure in the system (Table 9.5)

Table 9.5: *Problem: Low or no pressure in the system*

Possible cause	Remedy
Low setting of relief valve	Reset relief valve
Relief valve stuck open	Clean or replace
Leak in system	Check pressure drop
Broken pump parts	Replace broken parts
Incorrect control valve	Replace parts

Problem 6: Cylinder slow or not moving (Table 9.6)

Table 9.6: *Problem: Hydraulic cylinder slow or not moving*

Possible cause	**Remedy**
Directional valve failure	Check power input to solenoids
Insufficient pressure supplied	Check system pressure
Hydraulic line problem	Check for dented or crushed hoses
Defective actuator	Check piston rod bent or dented cylinder
Load exceeds capacity of actuator	Check system pressure and size of piston
Hydraulic circuit error	Check valve backwards
Air in system	Bleed air from system
Defective or worn pump	repair or replace

9.3 Maintenance of pneumatic systems

9.3.1 Maintenance parts of pneumatic systems

The main parts used for maintanance of pneumatic systems can be listed as follows:

1. **Coolers**

   As temperature of air increases during compression, removing heat during compression reduces the work required to raise the pressure of the air. Heat can be removed from the air compressor to the surrounding air or to water. Air-cooled compressors pass the hot lubricating oil from the compressor and compressed air through finned-tube heat exchangers and force ambient air across the heat exchangers using a cooling air fan. Cooling fan horsepower is typically about 5% of the power of the compressor motor. Water-cooled compressors use water-to-air heat exchangers to remove heat from the lubricating oil and compressed air. In many applications, this heat is eventually rejected to the atmosphere by a cooling tower. Increasing cooling by decreasing the temperature of the cooling air or water improves compressor efficiency and output capacity.

2. **Dryers**

   The two most common types of dryers for removing moisture from compressed air lines are refrigerated dryers and desiccant dryers. Dryers are typically sized to handle the peak air compressor air flow. As compressed air cools, water vapor can condense out of the air and should be removed from the compressed air system through drains.

3. **Filter, Regulator and Lubricator** (FRL)

   The filter removes particulates entrained in the compressed air and may have a trap or drain at the bottom. Regulator reduces downstream air pressure. Regulators have pressure gauges and valves to adjust the downstream pressure. Lubricator looks like the filter, but have a clear bubble or screw assembly on top for adding oil.

4. **Condensate drains**

   Condensate drains should be located:

- After the cooler.
- Underneath the receiver tank.
- At low points in the system.
- After filters, regulators and other devices that result in a large pressure drop.

9.3.2 Air treatment

Figure 9.1 shows a typical drawing of air filter indicating the different components required to exist in any air filter.

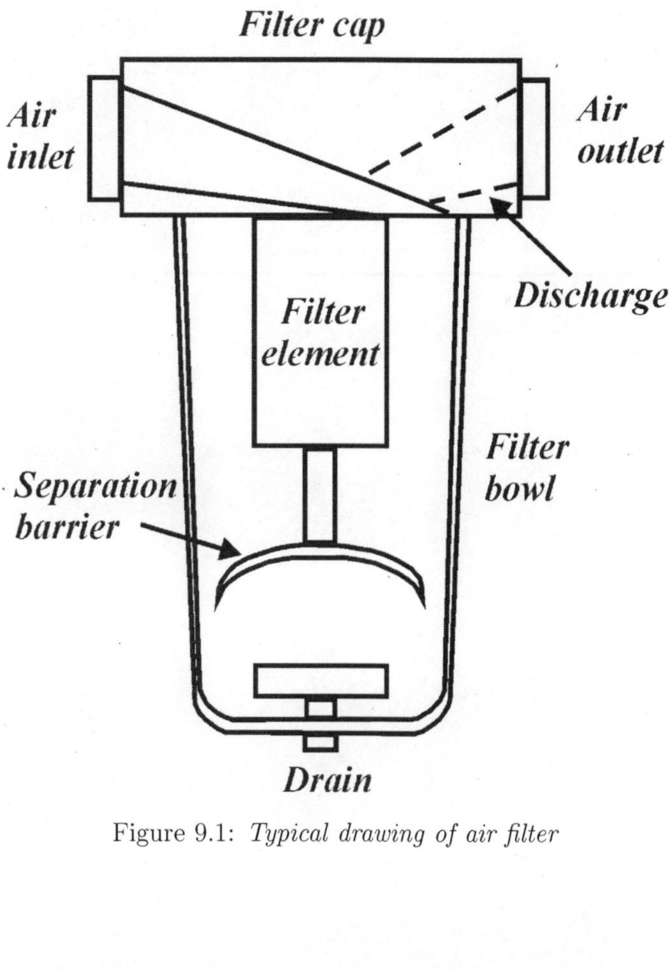

Figure 9.1: *Typical drawing of air filter*

Air filters can be calssified according to the filter element as follows:

- **Fiberglass filters**
  This throwaway air filter is the most common type. Layered fiberglass fibers are laid over each other to form the filter media and typically are reinforced with a metal grating that supports the fiberglass to prevent failure and collapse

- **Polyester filters**
  These filters are similar to fiberglass filters but typically have a higher resistance to airflow and a superior dust-stopping ability

Index